Printed in the United States
By Bookmasters

HIGHER EDUCATION DYNAMICS

VOLUME 11

SCOPE OF THE SERIES

Higher Education Dynamics is a bookseries intending to study adaptation processes and their outcomes in higher education at all relevant levels. In addition it wants to examine the way interactions between these levels affect adaptation processes. It aims at applying general social science concepts and theories as well as testing theories in the field of higher education research. It wants to do so in a manner that is of relevance to all those professionally involved in higher education, be it as ministers, policy-makers, politicians, institutional leaders or administrators, higher education researchers, members of the academic staff of universities and colleges, or students. It will include both mature and developing systems of higher education, covering public as well as private institutions.

The titles published in this series are listed at the end of this volume.

GOVERNMENT AND RESEARCH

Thirty Years of Evolution

by

MAURICE KOGAN

Brunel University, Uxbridge, U.K.

MARY HENKEL

Brunel University, Uxbridge, U.K.

and

STEVE HANNEY

Brunel University, Uxbridge, U.K.

A C.I.P. Catalogue record for this book is available from the Library of Congress.

ISBN-13 978-90-481-7130-9
ISBN-10 1-4020-4446-1 (e-book)
ISBN-13 978-1-4020-4446-5 (e-book)

Published by Springer,
P.O. Box 17, 3300 AA Dordrecht, The Netherlands.

www.springer.com

Printed on acid-free paper

Contents

Abbreviations vii

Introduction to Second Edition 1

PART I GOVERNMENT AND SCIENCE 7

1 Relationships between Government and Science 7

2 Theories of Science and Science Policies 23

3 Theories and Practice of Government 39

PART II THE COMMISSIONING SYSTEM IN ACTION 53

4 The DHSS and the Research Management System 55

5 Science and Macro Scientific Policy: the Case of the CSRC and the Intermediate Boards 61

6 The Chief Scientist's Organisation and the Research Councils: the Case of the Panel on Medical Research and Relationships with the SSRC 77

7 Research Liaison Groups and the Small Grants Committee: Two Contrasting Systems 93

8 The Chief Scientist's Organisation and External Research Bases: the Case of the DHSS Research Units 115

9 Review of Units and Scientific Merit: Chief Scientist's Visits 131

10 Review of Units and Policy Relevance: the Customer Review 153

PART III EMERGING PROCESSES AND ROLES 165

11 The Functions, Process and Impact of Research Commissioning 165

12 Emerging Roles 177

13 Policy after Rothschild and Generalisations 191

Appendix: Preface to the First Edition 213

Bibliography 219

Index 239

Abbreviations

ACSP	Advisory Council on Social Policy
AHPSR	Alliance for Health Policy and Systems Research
CBA	Cost Benefit Analysis
COHRED	Council on Health Research for Development
CPRS	Central Policy Review Staff
CRDC	Central Research and Development Committee
CSO	Chief Scientist's Organisation
CSP	Council for Scientific Policy
CSRC	Chief Scientist's Research Committee
DES	Department for Education and Science
DH	Department of Health
DHSS	Department of Health and Social Security
DSIR	Department of Scientific and Industrial Research
EAO	Economic Advisers' Office
EU	European Union
HPSS	Health and Personal Social Services
HSR	Health Services Research
HSRB	Health Services Research Board
HTA	Health Technology Assessment
MRC	Medical Research Council
NAO	National Audit Office
NEAT	New and Emerging Applications of Technology
NHS	National Health Service
NICE	National Institute for Health and Clinical Excellence
OCS	Office of the Chief Scientist
OECD	Organisation for Economic Co-operation and Development
OPCS	Office of Population Censuses and Surveys
PAC	Public Accounts Committee
PAR	Programme Analysis Review
PESC	Public Expenditure Survey Committee
PMR	Panel on Medical Research
PPBS	Programme, Planning and Budgeting System
PRP	Policy Research Programme
PSSRG	Personal Social Services Research Group

RAWP	Resource Allocation Working Party
R&D	Research and Development
RLG	Research Liaison Group
SDO	Service Delivery and Organisation
SGC	Small Grants Committee
SHHD	Scottish Home and Health Department
SSRC	Social Science Research Council
UGC	University Grants Committee
UKCRC	United Kingdom Clinical Research Collaboration
WHO	World Health Organisation

Introduction to Second Edition

Purpose and scope of the book

THERE HAS been a flare-up in interest in science policy, both in national governments and international bodies, and in the academic networks that track and criticise its progress. A key factor in this is the increased interest in analysing the role research can potentially play in informing policy-making. This is manifest, in particular, in areas such as health.

A pioneering venture in this field was *Government and Research: The Rothschild Experiment in a Government Department* (Kogan and Henkel, 1983).* This work, described in a sustained review in *Nature* as 'methodologically path-breaking', sought to depict the ways in which two sets of institutions, science and government, possessed their own characteristics which were however moulded and changed by the interactions between them. It sought to be an authoritative statement on the relationships between science and government and lodge itself in the political science literature of the subject. It thus fell into the tradition being established by American leaders in the field such as Caplan and Weiss.

It was a unique study, inasmuch as none other had penetrated the deepest recesses of government to observe at first hand the attempts of a major department - the then Department of Health and Social Security (DHSS) - to determine its research agenda through collaboration with leading scientists in a whole range of fields, to observe how research was commissioned, and then evaluated by scientific teams, and how it began to enter the policy blood streams of the departments. In order to do this, the two authors of the 1983 work had secured unrivalled access to private meetings and papers to the point of observing scientific groups being evaluated and the subsequent meetings and exchanges of papers within the Department. Over seven years it was possible to evaluate the whole cycle of policy into research commissioning and reception (See Appendix).

Much has changed since the 1970s and 1980s, but much remains the same. The forces at work in the story we told in 1983 about government and science have

* In the Appendix to this book we reproduce the Preface to the first edition, which contains an account of our methods and acknowledgements due to many collaborators.

grown stronger, if also more complex. 1971, the year in which the Rothschild Report was published, saw a major international shift in science policy, which has not been reversed. The idea that science, if left to itself, would serendipitously yield new discoveries that could be harnessed to societies' needs partly gave way to the view that governments, rather than scientists, should set research priorities and that social and economic goals should be the driver of science policies (OECD, 1971). The trend towards utilitarian goals and external influence on scientific agendas gathered momentum in the 1980s and 1990s. The 1993 UK White Paper on Science Policy made it clear that in future 'decisions on priorities for support [of science] should be much more clearly related to meeting the country's needs and enhancing [its] wealth-creating capacity' (para 3.9).

At the same time, industry became an increasingly important player, as collaborator with government and science in pursuit of market success driven by technological innovation, an idea that found expression in the Foresight policies adopted in a number of countries (Irvine and Martin, 1984; Martin, 1996). Long established boundaries not only between government and science but also between the state, the market and academia became more permeable, giving rise to a complex set of relationships sometimes referred to as 'the triple helix' (Etzkowitz and Leydesdorff, 1995, 2000) of government, business and universities.

'The knowledge society' has become one of the most universally adopted characterisations of the contemporary world, signalling, certainly, recognition of knowledge (not least science and technology) as a growing force in politics, economies and social organisation. Whilst the nature of knowledge (including science and technology) is increasingly contested, one of the most significant manifestations of its growing importance is the movement towards evidence-based approaches. A review of the growth of the evidence-based movement across a range of public services recently concluded that, 'the research community in healthcare is truly global, and the drive to evidence-based policy and practice is pandemic.' (Davies and Nutley, 2000). Within the UK, the National Audit Office (NAO) recently reported to Parliament on how government departments could best organise the commissioning of research so that it would inform policy: *Getting the Evidence: Using Research in Policy Making* (NAO, 2003).

Globalisation and internationalisation have also become increasingly prominent themes in science policies, even if many health and other researchers believe that international working best starts with good national systems from which individual researchers and groups can make their own connections. At the international level, some key themes relevant for our analysis were identified by the World Health Organisation (WHO) in a report prepared for the World Ministerial Summit on Health Research in November 2004. This World Report, *Knowledge for Better Health* (WHO, 2004), its conceptual framework for the analysis of health research systems (Pang et al., 2003), background papers (Hanney et al., 2003) and the work of an accompanying Task Force (Task Force on Health Systems Research, 2004; Lavis et al., 2004), all highlight the importance of organising national health research systems so that they can inform policies to improve national health systems.

Recognition of the desirability of undertaking research to meet the needs of potential customers in local health systems has resulted in much analysis of priority setting methods - in relation to both developing and developed countries (Global Forum for Health Research, 2002; Department of Health, 1993). The 'Linkage and Exchange' initiative developed by the Canadian Health Services Research Foundation is widely viewed as a significant model and involves bringing policy-makers who can use the results of a particular piece of research into its formulation and conduct (Lomas, 2000). Such a collaborative approach, or at least interaction between researchers and policy-makers, is increasingly seen as the way of producing research that is most likely to be utilised (Lavis et al., 2002; Innvær et al., 2002). Furthermore, there is a growing focus on the importance of brokerage or translator roles in the transfer of health research findings to policy-makers (Walt, 1994; Dash, 2003) and on the role of receptor bodies (Lomas, 1997; Hanney et al., 2003). Illustrating the greater attention being given to such ideas, several of them now feature in training packages about organising health research systems that have been developed under the Collaborative Training Program (CTP, 2004) by international bodies such as the Alliance for Health Policy and Systems Research (AHPSR) and the Council on Health Research for Development (COHRED). Finally, the notion of the customer for research is itself being expanded. Some governments have promoted an increased focus on the public's perspective in health research agenda setting (Oliver et al., 2004) in addition to more widely encouraging public engagement with health research and utilisation of its findings (Haines et al., 2004).

These developments have given rise to new theories about how research systems work, how knowledge is produced and how science-government relationships operate. However, they mostly reflect substantial continuities with those that underpinned our earlier study, which, because it covered a whole cycle, was able to identify the obstacles facing such moves as well as the potential benefits. We have been persuaded that the account given in our earlier analysis of theories and of developing government practice remain relevant 20 years on. This is attested by reference to it in recent discussions related to the above developments; for example, analysis of how best to organise health research systems and promote collaborative research (Denis and Lomas, 2003) notes a convergence between emerging forces within academia and changing norms within policy and management. It refers to the 'seminal' role of the Rothschild Report, stating that Kogan and Henkel 'describe the lessons from this era well'. It goes on to show how the emerging mode of collaborative research commissioning identified in the first edition of this book has now been bolstered by developments such as the increased interest in commissioning research that, as described above (Davies et al., 2000; NAO, 2003), is intended to lead to evidence-based policies.

We have kept much of the text of the original book, which remains a sustained case study provoking many themes still salient today. We have incorporated some new theoretical perspectives in Chapters 2 and 3. Otherwise the main changes come in the final chapter where developments since 1983 are drawn upon selectively and brought into the analysis. In particular, we describe how, in the 1990s, various strands from the Rothschild period were revisited by those responsible for the health research system in the UK in what was perhaps the first comprehensive attempt in any country to develop a national R&D infrastructure for the health care system (Peckham, 1999; Black, 1997). In drawing conclusions about the lessons from the Rothschild period, it has, therefore, been possible to illustrate their continuing relevance.

The structure of this book

The book retains its previous three main sections. Following this Introduction, Part I (Chapters 1-3) sets the context for the relationship between government and science by considering some of the relevant theories. Part II (Chapters 4-10) is the empirical heart of our study. Here we describe how the organisation of the DHSS (Chapter 4) was extended to include the research management and advisory committee system

(Chapters 5-7) and to attempt new relationships with the research councils. The particular case of the DHSS research units is taken up in Chapters 8-10 where we examine their purposes and the processes of 'peer' review by scientists and of customer review by the DHSS policy divisions. These chapters provide the empirical grist to our conceptual mill: they substantiate our theses of government's and science's multimodality and explain the difficulties of arranging fruitful encounters between them. In Part III (Chapters 11-13), we take up the same experiences to examine the processes, functions and outcomes of the research commissioning system and how it precipitated such new or reformulated roles as customers and receptors of research, brokers between science and government, and contractors attempting to meet government's needs. In Chapter 13 we also give a brief account of developments over the last 20 years using key points from our earlier 1983 edition to highlight some of the major advances and remaining problems. In drawing our generalisations and conclusions, we show how the concepts developed in the first edition are still of considerable relevance when attempting to evaluate and analyse recent developments in health research systems, and not only in the UK. We recall the methods employed in our seven year empirical study in the Appendix.

Prefatory note to the second edition

In bringing the text up to date, the two original authors are joined by Steve Hanney who has undertaken a series of studies in the field of health research systems over the last decade (for example, Buxton and Hanney, 1996; Hanney et al., 2000; Pang et al., 2003). All three authors are grateful to colleagues who encouraged us to undertake the second edition and provided expert advice, in particular, Martin Buxton, Robin Dowie, Shyama Kuruvilla and Bryony Soper. We are also indebted to Avril Cook who provided excellent secretarial assistance.

Maurice Kogan, Mary Henkel
and Steve Hanney, Summer 2005

Part I Government and Science

Chapter 1: Relationships between Government and Science

Our model

IN THESE chapters we recall the encounter between government and science in a stressful and volatile period of British political and social history. Its setting is one British government department, the Department of Health and Social Security. Much of what we have to say will be an historical account derived from our study of DHSS papers, from attendance at many of the key meetings, and from interviews and meetings with some of the principal actors. In the first half of Chapter 13 we bring the story up to date with an account of some of the significant changes in roles and relationships that have taken place since we published our first edition in 1983.

In order to wring the maximum benefit from the natural history of these events it is important, however, that we establish in the reader's mind the broad themes that underlie this history. We begin with the idealised and classic models of some key characteristics of two worlds - science and government - and their relationship with each other.

In the classic and 'internalist' model, science has its own structures of values and of knowledge. These constitute a complicated and varied world of their own. Science contains its own system of power and authority which underpins regulatory, allocative, rewarding and sanctioning institutions. Thus a member of the 'scientific community' is apprenticed to and becomes inheritor of the disciplined accumulation of knowledge, and of the rules by which knowledge is mastered, advanced, tested and refuted. These individual characteristics of the scientist respond to the broader relationships of power and authority applied by scientific disciplines as they license scientists and allocate them status and resources.

An equally idealised model of government assumes that it, too, has its value structures, most usually described in terms of bureaucracy and attendant managerial hierarchies and, somewhat less elaborated in the literature, its own structures of knowledge as well. Its knowledge systems have been typified in terms of the degrees of specialisation and generalism adhering to different functions and roles, and in some literature (Linder, 1980) in terms of the way in which objective data from outside become subjectivised within the system. Government has its own authority and power relationships encompassing the worlds of political direction, administrative execution of policy and client groups who form part of both the dependency relationship and groups which press upon the system producing statements of interest for government to reduce into allocations. And government, too, has its own institutions performing regulatory, allocative, rewarding and sanctioning functions.

The classic and idealised models of science and government both assume convergence and unity. They refer to somewhat autonomous entities brought into relationship with each other, from time to time and for particular purposes, but essentially capable of going their own way without decisive interpenetration or significant mutual effect. Science has been assumed to have its own resources and its own authority. The fact that much of the funding has come from government sources was not relevant because they were assumed to be grants made on relatively free terms. Equally, government was hardly challenged in its norm setting or in the preferences exercised between different sources of knowledge by its relationship to science. The notion of a free market of ideas in which government could choose among relatively self-confident providers seemed unassailable until the mid-1960s.

The simplicity of these classic assumptions has been drastically undermined during the period of our story. The elegant abstraction of the 'internalist' model of science (Merton, 1973) has given way first to epistemological doubt and then to sociological scepticism (Mulkay, 1979). Reflection about government, too, has become an arena in which behavioural scientists play with political, constitutional and organisational theorists.

What then were the modifications made to the presented model in our 1983 account? Certainly, we accepted that both science and government inhabit worlds of their own. But increasingly they had been pulled not only into each other's orbits but into those of other institutions too: boundaries were less distinct, and systems more complex.

Partly, but not only because of this, both science and government display dichotomous and conflicting characteristics of *convergence* and *divergence*. Science must act coherently. It can only, for example, license its practitioners through the award of doctorates, or be certain that a learned article is worthy of publication, or verify that an experiment has been duly controlled, or certify to the internal consistency and logic of an argument, if it is confident of its techniques of evaluation and of control and is able to insist upon reliable conformity to them.

There are, indeed, principles of testing, of accumulation of knowledge, of ways in which science might be refuted and augmented. Through power, authority and institutions it asserts degrees of uniformity and *convergence* of norms to which scientists conform. At the same time, however, *divergences,* less well acknowledged perhaps among the scientific elites responsible for the government of the scientific community, are evident. Much of the disciplined inquiry which the DHSS, albeit tentatively, sought to encourage, did not respond to classic notions of scientific control and had as its starting-point the problems set by client groups in the larger society rather than the unfinished business and logical imperatives of scientific disciplines. Moreover, even within those scientific areas where the client groups hardly penetrated, differences in criteria were evident. The norms set up by the sociologist, the anthropologist, the social psychologist are not the same as those of the experimental physicist, medical scientist or experimental psychologist which remained dominant. There is a wide range of epistemologies each of them carrying different normative assumptions about science. One of the fascinations of our study is the extent to which normative systems might have changed - but largely did not - in response to the pressures placed upon science by those acting as proxies for client groups in the wider society, and those demanding a different view of science in the scientific community itself.

But government too is tribal and divergent. The DHSS acted through many modes in response to the agenda set by ministers, by the dependent local and health authorities - and by client and pressure groups. In some modes it allocated resources and power and was closed and authoritarian. It must then be convergent. In other modes it reflects upon the aggregate of all of these things and tries to set them into priority order. In so doing it may be permeable by external influence. The judgements of its professionals are then modified by those of outside professionals and client groups. When it relates to the scientific community it may act as a customer quite instrumental in its demands or it may set itself up as a broker between customers and the scientific community. Even at the very top of the system in the 1970s and 1980s its different commands, led by Deputy Secretaries, the Chief Medical Officer, the Chief Social Work Adviser, grappled with issues from different dimensions. So did the Chief Scientist, a principal actor in our story - executor of, and adviser on scientific policy, and principal broker with the scientific community - work in different modes.

The *multimodlity* of government is evident whenever science and government engage each other. Each partner defines itself differently. Some brief examples may make this point clear. For example, when the DHSS was working, through its nursing divisions, on what disciplined inquiry will most help nursing develop, it was acting in a field virtually innocent of other sources of finance and also one in which it had a primary interest because nurses were almost wholly employed by the National Health Service for which the DHSS was accountable. The DHSS as the body commissioning research was in an entirely different position when it faced the Medical Research Council in the mode set for it by the Rothschild Report (Rothschild, 1971) - as the holder of funds transferred from the MRC (see Chapter 6), when it could invite the MRC to commission work which it deemed to be necessary. Both the MRC and the Royal College of Nursing defined their relationship with those who carry out scientific research quite differently from the relationship they had with government, or with other research councils, or with the manpower training system with which they were also concerned.

If the institutions of science and government change tactically in response to those with whom they must relate, so is their self-definition altered by the wider social context. The DHSS participated in the social and political movements of its time and thus went through a period of optimism and certainty about the role of government and its ability to commission knowledge that would be useful to it. This was later to give way to a more imperative mood, at the time of the installation of Rothschild, when it was no longer content to offer the blandishments of the market to researchers, but instead sought to be more systematic in its commissioning. And it later fell in with the mood of disengagement which emerged in government in the late 1970s with both a change of government and the general onset of pessimism about the power of rationality and knowledge to order human affairs.

We have, then, complex institutions of science and government. Each displays different modes of working and different degrees of certainty or convergence and volatility or divergence. Normative stances and institutional characteristics are changed and expand as each interacts with the other and with other institutions; at the same time each is resistant to pressure and change. Given these inherent characteristics, what relationships resulted from the commissioning of science by government? Our historical account disposes of the notion that the relationship between government and science has been, or can be, that of a simple managerial or hierarchical relationship, although much science is, of course, hierarchically and managerially ordained. Government departments have their own scientific units which, in differing degrees, are subordinate to the service needs of their departments. In general, however, and certainly in the cases to which we are referring here, there is a pattern of negotiation which needs other, and more complex, conceptualisations than those of naive managerialism. Here we followed the rediscovery of exchange theory already being so extensively applied to the relationship between central and local government (e.g. Ranson, 1980; Rhodes, 1979) and the corresponding consequences in terms of the distribution of power between different groups. A simple exchange model might assume that government gives resources for science in exchange for which scientists give their expertise and commitment to the solution of problems. As we shall see, however, this pattern of exchange, of mutual dependency, became strongly modified by the application of the customer-contractor principle. This wrongly assumed a simple exchange relationship in which the contractor could freely accept or reject government-commissioned work. Because of the conditions under which science began to operate in the mid-1970s, an imbalance of power, resulting from an imbalance in the terms of exchange, emerged, and patterns of negotiation gave way to other and less interactive operations of power.

A further and broader series of conceptualisations concerns the ways in which policy is made, and the ways in which science is called upon to contribute to that policy. It can be asked whether government's connections with groups outside itself are mediated through a system of consultations with elites within its domains of interest. This issue, classic in political science (Crewe, 1974), becomes important when we consider whether DHSS's commissioning is directed towards producing knowledge thought to be usable by those with most status in science, or whether it can accommodate wider groups of scientists, and client groups as well. Merely to introduce the full range of client groups, and a wider range of science, does not, of course, of itself entail pluralist models of decision-making. But there is a spectrum of such practices to which we can address ourselves.

In reciting the history of the DHSS's attempts to commission science before the Rothschild Report, during the high noon of the Rothschild pattern, and in the period when much of the Rothschild principle was dismantled, we shall be testing the assumptions contained in these paragraphs.

The working of each side can be described in terms of their range of values, epistemologies and institutional arrangements for mediating power and authority. As each seeks to work with the other side the resulting relationship can be described in terms of varying degrees of managerialism, negotiation, exchange and dependency. These relationships then take a particular form such as those of brokerage and, perhaps, loosely- coupled elites working within an essentially negotiative pattern.

The policy issues

The problems encountered by the government department and scientists are largely explained by the characteristics and relationships summarised in our critique of the classic model. The issues which the model should illuminate are, first, why is it so difficult for central government and the world of disciplined inquiry to collaborate even when, as in this case study, serious and responsible attempts were made to create conditions for that to happen? How far is government organised to use sources of knowledge outside itself? How far does the research community respond to government's encouragement to contribute towards solutions to the problems affecting public interest? What use can government make of their contributions and how does it evaluate them against those from other sources?

The policy problems may be answered at several levels. Taken as a whole, government may have a limited capacity to tolerate scientific inquiry that intensifies uncertainty or challenges its own working. Equally, science might be asked to meet the needs of society, or government, for information or conceptualisations of a kind that are not easily reconciled with its own structure of disciplines.

The levels of government at which definable and researchable problems can be identified may be limited. At the more baffling levels of policy formulation the scientist may fall away because the issues become predominantly those of values and allocation rather than of the discovery of fact, reanalysis of concept and the formulation of scientific conclusions. The question then arises whether there is such a thing as macro scientific policy. These difficulties become more real when we consider what affects the work of individual officials and scientists. The policy-maker, as we describe in Chapter 3, is the beneficiary and victim of several conflicting frameworks of knowledge and of social processes. Scientists strive to establish bounded frameworks with their own integrity and logic, and in such examples as those connected to the human genome or bio-technology have demonstrated the great power of science, but, as we contend in Chapter 2, they too respond to social influences beyond those of scientific norms.

The DHSS tried to overcome such problems by inviting academic advisers and researchers to collaborate with them in formulating research policies and implementing them. Why then did the mechanisms for ensuring co-operation become so laborious and eventually in part dismantled? Was the problem one of organisational structure? Or was the Department casualty to particular events and motivations? Or was the enterprise inherently impossible, because of the two different cultures which inhabit the worlds of governments and science (Caplan et al., 1975)?

The story

The story is a complex one and we briefly summarise it here. The Department had begun to commission substantial pieces of research throughout the 1960s. Particularly in areas of health and social services, programmes, sometimes incorporated in units (there were 38 of these, typically on six-year rolling contracts, by the beginning of the 1970s), were established. They were given considerable degrees of freedom to contribute, largely as they saw fit, to the development of a scientific community concerned with applications of particular relevance to policy and practice. The DHSS, in this 'golden age', seemed confident that the enlisting of science to solving problems could be achieved by relatively free negotiation in which customer interest was elicited but was not necessarily made decisive.

By the beginning of the 1970s, as we shall see from Chapter 3, both the optimism and the desire for certainty became stronger. With the return of Edward Heath's government in 1970, government's search for knowledge and rationality became more pressing and the concepts of the client group within government and the customer in the policy division became more strongly sponsored. These trends were visible throughout government, not only in the DHSS, and the most important result was the Rothschild Report which, together with the Dainton Report* was published in 1971 as a consultative document. Almost immediately the government formally accepted the idea of the customer-contractor principle. A White Paper in July 1972 essentially followed the Rothschild Report and modified only the percentages of the research council budgets transferred to government departments.

What Rothschild said

The Rothschild Report made a sharp distinction between fundamental and applied research. Basic research, the province of universities and research councils, is research aimed at furthering 'discovery of rational correlations and principles' (Rothschild, 1972) while applied R & D has 'a practical application as its objective'. For 'applied' research to be funded it must have a named customer, 'the customer says what he wants; the contractor does it (if he can); and the customer pays.' (Ibid, para 6) All applied research funded by government departments should be organised on this principle.

Applied research is distinguished by its objectives. It is not defined in terms of the length of time it takes nor of the techniques or kinds of science it employs. To those who might argue that such a sharp distinction between basic and applied research ignored the potential interactions and spin-offs from each other, and that scientists themselves were capable of identifying social objectives for research, Rothschild replied 'the country's needs are not so trivial as to be left to the mercies of a form of scientific roulette, with many more than the conventional 37 numbers on which the ball may land.' (ibid, para. 6). He might have added that the distinction was intended to settle managerial accountability and public policy issues. It could not settle the ways in which science might organise itself.

Three principal recommendations for government departments followed. First, the customer-contractor principle should govern all applied research. Second, each

* *The Future of the Research Council System,* Report of a Council for Science Policy Working Group under the chairmanship of Sir Frederick Dainton, was published as part of the same document as the Rothschild Report. It rejected a distinction between pure and applied research because of the interdependence of the two for progress in each, and because the blurring of boundaries between different scientific fields leads to more internal cohesion of science. Instead it proposed a threefold classification of scientific work: *tactical* - that needed by government and industry to further its immediate concerns, whether research involved was long-term or short; *strategic* - general scientific knowledge underlying tactical science; and *basic* - research and training with no practical objectives other than advancing scientific knowledge and maintaining a corps of trained scientists. Its main recommendations were that the five research councils should continue to function as they were, but that the Council for Scientific Policy should be replaced by a Board of the Research Councils.

government department funding research was to appoint both a Chief Scientist to advise customers on research needs, and a named controller of research and development to be the executive head of the R & D function and to provide that service for the customer through either in-house facilities or external commissions. Third, varying percentages of the budgets of the Medical Research Council, the Natural Environment Research Council and the Agricultural Research Council were to be transferred to the relevant government departments in recognition of the applied nature of some of their work.

The control and transfer of funds from three of the research councils would ensure that government departments got what they wanted from them. The proportions of funds taken over by the departments were estimated from the amount of applied research sponsored by the research councils.

The 1972 White Paper was a child of its time. It was optimistic about government's ability to think and act. Thus:

> The new framework provides a partnership within which science will have more influence on the government's central policy-making' activities than before, and which will contribute more directly and more effectively to the task of making the best use of science and technology for the needs of the community as a whole (para 61).

The government hoped that application of the customer-contractor doctrine to all of its applied research and development would create clear responsibilities. 'Departments as customers, define their requirements; contractors advise on the feasibility of meeting them and undertake the work; and the arrangements between them must ensure that the objectives remain attainable within reasonable costs.' (1972 White Paper, para 61).

In the light of later experience, the Rothschild formula can be criticised for assuming that government departments were the only source of policy development, that they could state all their requirements from their own sources of knowledge and problem-setting. It failed to note how in those areas of policy where data are diffuse, and analyses most likely to be strongly influenced by value preferences, problems must be identified collaboratively between policy-maker and scientist. It failed to acknowledge that policy-makers have to work hard to identify problems, to specify research that might help solve them, and to receive and use the results of research.

Equally, it assumed that research institutions were strong enough to negotiate with government in a market where science might be procured as piecework. The need to defend a science system whose resources and legitimacy were already becoming attenuated, and which were to come under increasing threat as the 1970s progressed, was never anticipated by Rothschild. It assumed that science was sufficiently developed to be used, and thus failed to specify the need for government to assist in the development of science more closely related to practice and service needs than to academic disciplines. It could not anticipate the ending of the UGC quinquennial system or the reduction of research council monies. Possibly the customer-contractor relationship has a better chance of success when the product of research can be more easily specified. But it can work in the social policy areas on the assumptions set for it only if there is much more sensitive elaboration.

Reactions to Rothschild

The reaction of scientists was immediate, almost wholly hostile, but limited to concern for the effect on the research councils (eg *Minerva*, 1972). They argued that Rothschild provided no evidence that the research councils had failed in their tasks. They criticised as artificial the assumed divide between fundamental and applied research, and emphasised the stimulation that each could provide to the other. Moreover, the government departments would not necessarily be more successful in predicting future needs than researchers themselves.

But the issue was not whether science should be free or controlled, but what balance should be struck between the two modes. Science had become so large-scale an investment in many areas where government's interest was strong that the policy-machine could not remain detached from it. Moreover, not all scientists wanted to stay aloof. Some thought it important to have an impact on policy even if it meant adapting their research objectives to customer wishes. That did not mean, however, that all would accept that policy-makers should set the goals for science in the way the Rothschild formula prescribed. The relationship would involve negotiation of *quid pro quos*.

Despite scientists' objections, and these came mainly from those strongest in the academic setting rather than those primarily concerned with application to policy and practice, the DHSS acted upon the terms of the White Paper. It took up membership of the Medical Research Council, where previously the Chief Medical Officer was an assessor only. And the Department established an elaborate and multi-level set of committees to draw on the advice of established scientists. These were to give scientific credence to the research programme, and to build connections between policy and science.

The structure of the commissioning system

The DHSS adopted its own version of the Rothschild structure (see Figure 1.1). As noted earlier, Rothschild recommended departments to create two distinct roles: an advisory Chief Scientist to assist customers, and an executive Controller R & D, primarily to provide the customers with an efficient R & D service.

Until 1982 no one at the DHSS was designated as the Controller R & D. In the period 1972 to 1978 the Chief Scientist (Figure 1.1 (vi)) had an advisory function. This involved him in the selection of scientific advisers, the appointment of members of his advisory committee system (Figure 1.1 (i) - (v)) and the assessment of the quality of research, often delegated to advisers and research managers, although the Chief Scientist led the reviews of the DHSS-funded units. Executive action was undertaken by the research management division led by an Under-Secretary. That division was accountable for the research budget and the management of resources was the province of its administrative staff. Research management was shared by career administrators and by professional staff from the medical and nursing divisions and the social work service.

The Chief Scientist's Committees

The DHSS created a committee structure which should have ensured negotiation and collaboration between the Department and the scientific community at all levels.

The Chief Scientist's Research Committee (CSRC) (Figure 1.1 (i)) was to concern itself with all aspects of DHSS-funded research in health and personal social services. It was not able, in its relatively short life, to cover equally all DHSS concerns, and thus worked to a far lesser extent with social security and the more specialised research programmes of computers, supplies and equipment in buildings. The CSRC was to assess the priorities within the entire research programme, to ensure achievement of scientific standards and to consider the adequacy of resources for research. Its twenty members were drawn mostly from social medicine health service studies and the social sciences, reflecting

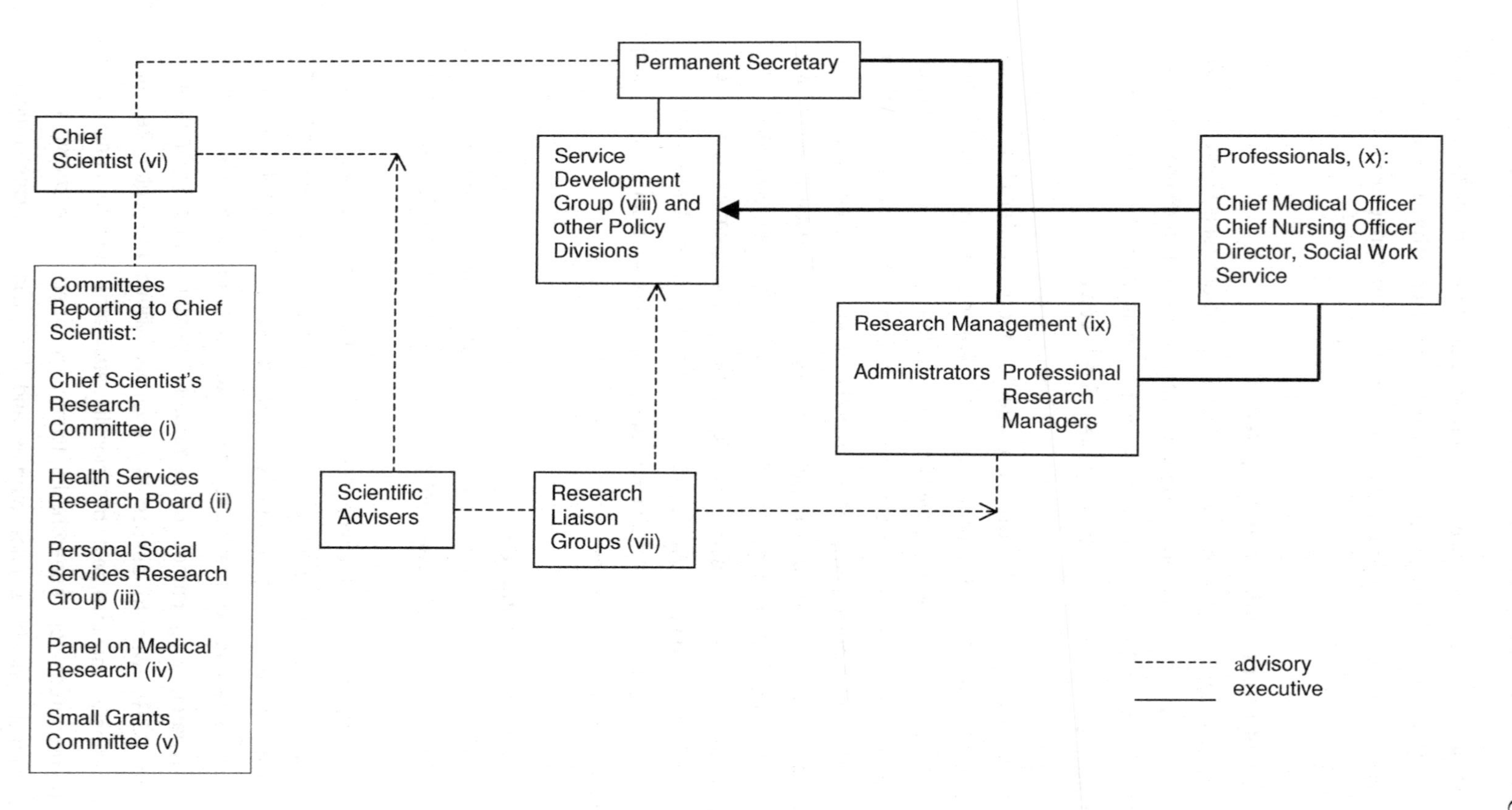

Figure 1.1 Structure of the DHSS commissioning system c. 1974

its primary concern with health and personal social services research. Two members were drawn from the field services. The CSRC and the other committees created at the same time reported to the Chief Scientist. (Figure 1.1 (vi)).

Two intermediate committees, the Health Services Research Board (HSRB) (Figure 1.1 (ii)) and the Personal Social Services Research Group (PSSRG) (Figure 1.1 (iii)) were expected to perform similar functions to those of the CSRC within their subject areas. The great majority of members were scientists. Practitioner and policy inputs remained relatively weak. The Panel on Medical Research (PMR) (Figure 1.1 (iv)), acted in parallel with the intermediate committees. Its function was to advise the Health Departments on the commissioning of research from the MRC, £9 million of whose funds were transferred to the Department under a concordat signed in 1973 by officers of the DHSS and the MRC.

But perhaps the most important move towards providing a forum in which the customer-contractor principle would be acted out was the creation of Research Liaison Groups (RLGs) (Figure 1.1 (vii)) which brought together the policy divisions and scientific advisers to work out research programmes for specific policy areas. They were chaired by a senior officer from the policy division - the customers - and contained external scientific advisers as well as policy-makers, DHSS professionals, and research management representatives. Their task was to identify the research needs of the policy divisions, to develop research programmes to meet those needs, to monitor research in progress and to review and act on research results.

As a complement to the Rothschild principles, a Small Grants Committee (SGC) (Figure 1.1 (v)) was to deal with proposals submitted by researchers spontaneously. The Department wanted still to respond to ideas from the outside provided they fell within the Department's areas of competence.

Thus from an age of relatively free negotiation, in which departmental research was encouraged largely through the work of creative individualists amongst administrators and professionals, there was a shift, following Rothschild, towards systematised collaboration: between the Department, the research community and later the field authorities, in pursuit of the customer-contractor principle. This entailed creating committees to determine a cross-departmental research strategy, to encourage substantive dialogue and decision-making between the Department and the research councils, to identify the research needs of individual policy areas and to commission work to meet those needs. From such a structure emerged a set of complex *brokerage* roles; the Chief Scientist and research managers, some of whom acted as liaison officers between policy-makers and researchers, needed to penetrate the workings of both the policy and the research system if they were to establish terms of exchange under which scientists could usefully contribute to policy. The stage was set for a long-term enterprise.

By 1978 another set of influences was on the move. Advisory and executive functions were now combined in the Chief Scientist's role to which Professor Arthur Buller was appointed. He assumed responsibility for the research programme and the central precept of the organisation became scientific accountability. By the end of 1978 the elaborate structure of committees had been largely dismantled. Only the RLGs and the Small Grants Committee remained, and there were soon to be 'plans to reduce activity. . . tailored to meet the circumstances of each RLG.' (CSRC (79)/2). It was recognised that the Department had failed to assess research needs in important non-RLG areas. Adequate machinery with appropriate scientific advice was felt to be lacking for identifying research needs in such areas as hospital acute services, primary health care, dental health, NHS manpower and personnel, hospital non-clinical support services, public and environmental health, medicine and central planning and resource allocations. But the trend that set in was towards attenuation not extension of existing machinery.

A rigorous review of the DHSS-funded research units was begun in 1979 aimed at raising scientific standards of DHSS-funded research but also at reducing the proportion of the budget allocated to this resource and now felt to be excessive. This led to reduction in the Department's capacity for Health and Personal Social Services (HPSS) research. Conversely, the position of the MRC was enhanced. In 1981 the funds transferred to the Department from the MRC budget were restored to the Council under a new concordat. In return, the MRC undertook to expand the scope of its work with a view to establishing a base for health services research, a move bitterly opposed by some leading researchers in that field. One academic who advocated the creation of a separate health services research council, anticipating the eventual decision, considered there to be 'an enormous chasm between the kind of work for which the MRC is rightly renowned and the kind of work wanted ... why lock [such research] into the MRC, where it will inevitably find itself dominated by medical thinking and attitudes and expected to conform to criteria flowing therefrom?' (Williams, 1981). In the event, the MRC initially set up a Health Services Advisory Panel, not a board, upon which there was minority social science representation.

The Departmental Research Strategy Committee set up following the demise of the CSRC had no external scientific representation. The new Chief Scientist appointed in 1982 was to work part-time, although an additional post of Deputy Chief Scientist was created to fulfil the function of Controller Research and Development.

This summary of events displays the difficulties encountered by the DHSS in responding to the strong and simple model of science-government relationships advocated by Rothschild. The events will be examined in more detail in later chapters of this work; where we will also bring the story up to the present time. But first we shall explore the developing notions of science and government that formed part of the context of government-science policies during our period. The influence upon the events described of the far-reaching political and economic changes in that period is undeniable and not easily separable from the changes to be marked in the two systems. But we seek to show that an explanation of the events can also be found in the workings and inter-relationships of those systems, once the apparently simple model from which we began is taken apart.

Chapter 2: Theories of Science and Science Policies

WE NOW examine more closely some of the characteristics of science and the members of the community who provide it, as they have been portrayed by scientists themselves and by theorists from various traditions. Several linked debates about the nature of scientific knowledge and its organization are particularly relevant to our story. The first concerns the extent to which science can be understood as a closed, self-regulating institution producing a distinctive form or forms of knowledge. Second, is the question whether science is a unified, convergent system. The third concerns how the aims and practices of scientists are most convincingly explained. All are connected to debates about how far science can or should be open to external influence, negotiation or control. Thus, what we have to say about these issues directly flows into what we have to say about government's 'steerage' of science.

We shall suggest that scientists themselves have moved from assuming that scientific activity is all of one piece. It involves many different philosophies and modes of verification. For this reason, some authors on the subject use descriptive terminology such as 'disciplined inquiry' or 'research and development' as well as 'science'. Because science is various and its boundaries the object of subjective appreciation it participates in many different relationships with society at large and its financial sponsors in particular. The DHSS, it will be shown, were concerned with multiple forms of scientific activity.

Theories of science

'Internalist' models of science have certainly exerted a powerful influence not only upon scientists but on those who have admiringly observed science's growth and strength. In the 'internalist' view, science is an authoritative and self-regulating institution. The nature of scientific work, its evaluative criteria, its institutional norms and structures are regarded as logically connected and rooted in the relationship between science and the physical world. The goals of science are 'the extension of certified knowledge' (Merton, 1973). Science uncovers the regularities of nature through theory generation, accurate observation and empirical testing. It expresses and explains them in laws that are both as precise and as general as possible. The criteria of scientific merit are thus accuracy of observation and measurement, replicability of experimental work entailing rigour in design and control, validity and systematic importance or profundity of theory. The derivative and tightly interconnected technical and moral norms of logical consistency and impartiality are strongly embedded in Merton's classic statement of the four sets of 'institutional imperatives' embodying the ethos of modern science: universalism, communism (sic), disinterestedness and organised scepticism (ibid.).

The intemalist models of science are, however, by no means uniform. While for many scientists objectivity is one of science's prime characteristics, Polanyi (1958), for example, rejects as sterile the categorisation of knowledge as subjective and objective, and welcomes the rapprochement in twentieth-century science between empiricism and theory. He recognises the powerful tension in the history of science between originality, creativity and profundity, and accuracy and reliability. He sees the key to their resolution in, first of all, the individual scientist, whose most distinctive quality is not his detachment but his 'heuristic passion'. Then the validity of scientists' work is enforced not by objective proof but by the exercise of responsible judgement. Because of the tension between scientific values, scientific judgement cannot be analysed; it is mediated through the authority of the scientific community. It is, in his view, essential to sustaining the commitment of the individual scientist and the authority of the scientific community that science remains self-sufficient with the freedom to push its norms to their limit. Its authority then is such that in a liberal society it can impose 'the intrinsic value of the scientific process on society at large' (Polanyi, 1962).

The common themes in these views of science are its unity, its impermeability and its authority. The themes of unity and impermeability find strong echoes in the evidence given to the Public Accounts Committee (PAC) in 1979 by the secretary of the MRC (PAC, 1979) and have affected the views of many prestigious scientists on the ability of government to promote research. They were given more flexible expression in the Dainton Report, but the theme of unity was unequivocally endorsed. 'The internal cohesion of science is one of its most characteristic features and will surely increase rather than diminish.' (Dainton Report, 1971).

There are differences of view about the basis of the authority of science. As indicated above, in one view the emphasis is upon its objectivity and impartiality while in another it is upon passionate commitment to the pursuit of truth. The prime benefit from the first is the provision of impartial evidence while that of the second may be creative theory, the use of which is not predictable. The over-riding importance of what goes into scientific work is stressed too by Popper: the unique authority of science is rooted not in its outcome but in its method of putting its propositions to the test. The 'logic of scientific discovery' insists that knowledge is provisional. Theories are to be judged not by their approximation to a (for him) unattainable truth but by their audacity, their scope, their explanatory power and their falsifiability. Also 'the objectivity of scientific statements lies in the fact that they can be intersubjectively tested.' (Popper, 1972). Popper's conception of science is thus far from depicting an uninterrupted accumulation of knowledge. For him, its strength derives from the willingness of its practitioners to place their theories and their life's work permanently at risk (Magee, 1973).

The thesis that most famously challenged Popper's version of the internalist model of science and its form of functionalism was that of Kuhn (1962; see also Kuhn, 1974). His picture of 'normal science' is one in which scientists see their task as puzzle-solving within the paradigms of their chosen disciplines; of a profession working within a set of established theories rather than seeking to test or challenge them. Within this view science grows or makes progress only through periodic revolutions forced through when those paradigms can no longer accommodate the perceptions and problems with which they are confronted. Moreover, paradigms are determined not simply cognitively but socially by the community of the particular discipline in which individuals work and as such they die hard.

Two consequences follow. One is that science can be seen as a social phenomenon, a permeable institution. Although Kuhn himself did not explore the social processes at work in science, he opened up the way for sociologists to do so and to go on to explore the relationship between the cognitive and institutional structures of science and between those and its socio-political environment (Mussachia, 1979). The second consequence is that science's claim to be a distinctive and pre-eminent form of knowledge is undermined. It is a possible, although not necessary, step from Kuhn's thesis (although not one taken by him) to consider science as one of a number of epistemologies with equal claims to validity.

The 1970s saw a number of challenges to Mertonian functionalist explanations of the nature and authority of scientific knowledge and the role in them of accepted scientific norms. Some pointed out inconsistencies between Merton's account of the ethos of science and both scientists' behaviour and the workings of the institution of science (Mulkay, 1977). Some suggested that scientists operate under conditions of normative ambivalence (Mitroff, 1974; see also Gornitzka, 2003). Our own more recent studies lend support to this interpretation: they suggest that traditional

Mertonian norms retain a strong place in the identities of scientists, even as they find themselves increasingly implicated in the norms of a market environment (Henkel et al., 2000; Henkel, 2005). Meanwhile some of those working in the Mertonian tradition argued that norms, rather than guiding scientific practice, were effectively used as rhetorical strategies by scientists to retain their status and authority in society (see Lynch, 1993).

Mulkay (op.cit.) took up questions of structure, power and ethos within the institution of science. He argued that the prestige structures, reward systems and endorsement of developments in science were not based solely on these norms. They were not even wholly consistent with them. He contended that recognition by the profession was the key objective of scientists and that this, in effect, meant recognition by the scientific elite. The 'scientific community' was, *pace* Polanyi (1962) not a republic; it was, rather, a complex nexus of problem-focused, discipline-centred and wider networks of elites able to perpetuate themselves through interaction between differential allocation of resources, differential capacity to recruit the best talent, and a privileged informal communication system. The norm of communal knowledge was undermined by intense competition for recognition. The race to be first depended upon secrecy, or at least highly controlled communication (cf. Watson, 1968), and the scales were weighted towards recognition through conformity to established concepts and techniques. Weingart's work (Weingart, 1974) on the institutionalisation of paradigms and their subsequent domination of the cognitive orientation and resource allocation in particular fields had also suggested interaction between the formation of elites and conservatism in science.

Whitley's comparison of restricted and unrestricted (see also Pantin, 1968) or configurational science made a similar point, but aimed to show further how the cognitive structures of different sciences give rise to different forms of organisation, and so to different degrees of cohesion and power (Whitley, 1977). He contended that the arithmetical ideal and the aim of expressing theory in closed and simple formulations inhibit challenge in restricted sciences such as physics, concerned with a small number of properties of objects, which can be quantitatively related. The high degree of specialisation needed creates clear boundaries within these sciences, bureaucratisation in the organisation of research and success in attracting resources. Configurational sciences, such as social sciences concerned with small numbers of highly-structured entities exhibiting a large number of properties, are essentially poly-paradigmatic. 'Attempts at monistic reductions [are] fundamentally mistaken.' Indeed, clearly distinct paradigms are not a feature of configurational sciences; rather they exhibit 'a number of inter-related approaches which diverge on some points and overlap on others.' Their conceptual boundaries are 'highly fluid and permeable' (ibid.). In consequence their organisation is less structured and there is greater scope for dispute and fundamental challenge. This in turn affects their authority outside their boundaries.

Other approaches to the study of science could be seen as a resurgence of the 'externalist' view of science that had briefly flowered in the hands of Bernal and others in the 1930s (Bernal, 1939). Then it had been essentially a Marxist perspective, which perceived the growth of modern science as rooted in the socio-economic organisation of society, its direction and specialisations deriving from the technical problems facing newly industrialised nations. Now, rejection of the view that the development of science could properly be understood only by reference to its own internal dynamic was taken up from a broader ideological and disciplinary perspective. There was interest in a wider range of influences on science, gender, for example, as well as class and power (Gornitztka, 2003). The architects of the British 'strong' programme contended that it was possible to explain the directions taken by scientific knowledge by reference to particular social or political interests among scientific elites that 'predisposed an acceptance or rejection of one or another possible theoretical interpretation' (Lynch, 1993). Further, while they, unlike some of the other more radical theorists, accepted that there is a boundary between science and society, they regarded it as a social construction. Their agenda was 'no longer to examine the social influences that operate across the boundary but to examine how the boundary itself is a product of the social organisation of scientific activity.' (Barnes quoted in Lynch, op.cit.).

Overall, it is clear that there have been a number of convincing challenges to conceptions of science as a unified social institution, driven by a unique set of norms and the logic of its own discoveries, and insulated from the kind of self-serving instrumental rationality assumed to be embedded in other sectors of society. In consequence, at the least, the boundary between scientists and non-scientists can be seen as less rigid and more open than suggested by the internalist model. Since the time of our study, social theorists and theorists of science have suggested more radical changes in the relationship between science and society. By 2001 Nowotny et al. were proposing that science has come to be practised in a diffuse and unbounded context of social implications. In these circumstances, they suggested, science becomes open to the involvement of increasingly heterogeneous populations. Society 'speaks back' to science, now operating in the '*agora*'. Science is no longer mediated and regulated only through a limited number of closed academic, professional or bureaucratic institutions, but also engages in collaborations, negotiations, debates and conflicts with all sorts of actors.

In the next section, we shall look more closely at aspects of the interest among theorists of science in patterns of relationships between scientists and other actors. In the process we consider some assumptions about the relationship between fundamental or basic and applied research and between science and technology that were challenged during the period of our study and afterwards. We are following the example of those who reject the polarities of the internalist/externalist debate and believe instead that 'science is complex rather than coherent in both its overall conceptual and social structure' (Krohn, 1977). In particular, we look at theories and conceptualisations that open up the complexities of the production and use of scientific knowledge and in so doing provide some tools for the analysis of assumptions about the possibilities of the steerage of science by governments.

Steerage and the fundamental/applied dichotomy

The separation of fundamental and applied research within a linear relationship underlies the Rothschild principle that customers will formulate the problems and contract them out to scientists applying basic knowledge. Theorists of science suggest that such a clear statement of relationships and tasks is over simplified. Some argue that there are emerging new modes of knowledge production that signal a 'deinstitutionalisation' of science and a fundamental change in its cognitive and social structures. According to Mendelsohn (1979) the boundaries established round science during the eighteenth and nineteenth centuries are 'breaking down as scientists and non-scientists re-examine the relationship of science and social needs, problems and values and as external factors increasingly steer research.' (in Mussachia, 1979). Theorists of the relationships between science, technology and innovation argue that the transfer time between basic research and technologies 'has been so far reduced that the institutional distinction between the context of (academic) research and the (non-academic) context of application has become obsolete in organisational terms' (Weingart, 1997). Linear models need to be replaced by models that can accommodate multiple 'patterns of feedback and recursivity' (ibid.; cf. Martin and Nightingale, 2000) as academic scientists from a variety of disciplines collaborate with different industries to develop new technologies.

An influential formulation of some of these developments was put forward by Gibbons et al. (1994) in their contention that towards the end of the 20th century the mode of production of knowledge could be understood as undergoing a radical shift from Mode 1 to Mode 2. Mode 1 knowledge is 'generated within a disciplinary, primarily cognitive, context', in which problems, rules and evaluative criteria are internally determined. Mode 2 entails a broader conception of transdisciplinary knowledge, generated within a context of application. It addresses problems identified within a context of continual negotiation between actors from a variety of settings.

Critics of Gibbons and his colleagues (see also Nowotny et al., 2001) argue that a polarised framework, in which two modes of knowledge production are set against each other is too simplistic. It fails to take account of the importance of contexts of application in the past (Godin, 1998; Rip, 2000). Many new locations of knowledge (such as think tanks and consultancy firms) remain largely dependent on disciplinary research in terms of knowledge flow and training and tend to operate in the 'transfer' sphere rather than in original research (Weingart, ibid.). Recent case studies of health services research in the UK suggest it involves a mix of Mode 1 and Mode 2 production of knowledge and that 'the basic disciplines retain greater defensive power than Gibbons et al. suggested' (Ferlie and Wood, 2003).

As early as the 1970s, Weingart and his colleagues had begun to analyse the implications of installing strategic direction of science in the name of social problem solving (van den Daele et al., 1977). They argued that the conversion of political into scientific problems could be significantly influenced by administrative structures and interests and also by the existing (disciplinary) institutionalisation of science. They would determine who is selected as scientific experts and how they perceive problems and the search for solutions. They further demonstrated how finalisation theories of science could illuminate the variety and complexity of the possibilities of political direction or steerage.

They identified three phases of discipline development: the exploratory, pre- or polyparadigmatic phase, the phase of paradigm articulation and the post-paradigmatic phase. They went on to assert that in the first and third phases problem orientation and discipline development are compatible. Before paradigms are established functional research can at the same time be an input into the discipline. But at the point where work is beginning to crystallise into the development of key theoretical models, usually the research programme is dictated by 'internal' needs incompatible with external problems. At the third stage, when the basic explanatory models have been tested it may be possible to coordinate their elaborations with problem-oriented research. From this time onwards such research may instead be carried out by application of the basic theory.

Many research fields of importance to the DHSS, such as health services research, general medical practice, social work and nursing, could be understood as in a pre-paradigmatic or exploratory stage, at least at the time of our study. There were no established or dominant paradigms, no agreements on the most relevant disciplines, and no clear academic networks. Moreover, the impetus for their development may have been as much practical as scientific: those prominent in some of them might not regard themselves primarily as scientists or even as researchers. Theoretically, researchers in these areas might be susceptible to steerage by government, but at their stage of development it might be difficult for either them or government to formulate clearly researchable questions. Their relative lack of clear reference groups might make it hard for them to demonstrate the distinctiveness of their contribution if government sought to impose its own frameworks on them at too early a stage.

Alternatively, an established science of recognised importance to the Department might be at a stage in which the old paradigm is under question. Thus at the time of our study epidemiology seemed to raise a double problem. It was moving from a position in which the nature of evidence and measurement and the choice of methodologies were undisputed into one where, because social and psychological factors and qualitative rather than quantitative questions were assuming dominance, this was no longer the case. Moreover, this meant, too, that the value issues were coming to the surface and therefore ends/means relationships and criteria of effectiveness were by no means self-evident or even relevant.

Sometimes problem orientation can produce its own coherent research community or institution, although more often the key problems of directed research are taken up by 'hybrid communities', multidisciplinary institutions incorporating non-scientists as well as scientists. Sometimes these, too, may formally resemble the classical scientific communities and develop their own evaluation standards, reputation structures and career patterns, although knowledge is produced according to explicitly identified social or professional norms. However, in some of the cases investigated by 'finalisation' theorists theoretical development (not simply application) was necessary for the resolution of problems emanating from the policy fields. Again, sometimes the need for new theoretical models to reconstruct policy problems into scientific ones might mean intra-disciplinary development, even for 'hybrid communities'. In other cases, such as environmental studies, it meant theoretical formulation within new constructions of the relationship between humans and their environment (Weingart, 1977; cf Buxton, 1981).

In this literature, it seems that simple distinctions between fundamental and applied research may be of limited use as the complexity of policy problems, the rate of change and the demands upon government to respond increase. This view in some respects accords with that of Trist (1972) who in the context of social science strongly challenges the assumption that concern with the advancement of knowledge is the monopoly of fundamental research.

His formulation is particularly important to our developing argument. In positing three forms of scientific activity - basic, applied and domain-based research - he outlines how all three advance knowledge. There is the development from discipline-based theory and method in fundamental research; the nurturing of often unfashionable innovations in service or profession-oriented applied research; the identification of meta-problems in domain-based, policy-oriented research which is essentially inter-disciplinary and where the crossing of new boundaries and creation of new syntheses may advance both knowledge and human betterment on a new level of comprehensiveness. Indeed, he goes further, contending that the problems of access for social scientists make engagement with practitioners and policy-makers in problem solving, 'a fundamental strategy for advancing the knowledge base' (ibid.). An example in our study can be found in the work done with Down's syndrome babies and their parents at the Hester Adrian Research Centre in Manchester. There an experimental project giving service to these families had as an integrally linked objective fundamental research into the development of motor skills in the babies. The fundamental-applied research distinction had no practical meaning in this context and there was no sense of a linear relationship between the two.

In the face of turbulence and complexity distinctions between ends and means, and between fundamental and applied research, become less simple. So too does the relationship between natural and social science.

Natural and social sciences

A concern to understand and to appraise the relative characteristics and claims of the natural and social sciences is crucial for a department concerned with health and social services. Changes in concepts of health, disease and what constitutes health care have reflected a belief that physical, psychological and social processes are connected, but in a complex way. Medicine has been defined as 'applied social science'. It was a long time before it became linked with natural science and, indeed, has not done so in some countries. Yet, for the most part, medicine is now firmly rooted in the natural sciences and can demonstrate the advances that have accrued from working within that tradition. By contrast, social science is constantly under pressure to demonstrate the value of its contribution. The relationship between the two kinds of science is complex, too, and has not been clarified within the policy system.

Although the natural sciences dominate concepts of science at large, the attitudes of many social scientists towards them have long been ambivalent. 'On the one hand there is a powerful aspiration for a body of "scientific" knowledge of social life. On the other, there also exist strong doubts as to whether social life is ... a subject matter suited to the physical science model.' (Lessnoff, 1973).

The positivists' search for regularities, systematic explanation and prediction in social life enormously extended our perspective on the social world. But the pursuit of the values of the natural sciences in the social sciences, namely accuracy, replicability, rigour in experimentation, laws that combine a maximum of generality with a maximum of precision, inevitably causes the social sciences to be regarded as underdeveloped and inferior, because they cannot convert social complexity into analogues with the natural world. The tensions embedded in scientific work between the rigorous and the creative values will be much greater for social scientists. People are not simply 'objects whose behaviour is in principle explicable in terms of a series of natural laws.' (Becher and Kogan, 1980). The concepts of intention, meaning and value are central to an understanding of human action and a grasp of them entails a comprehension of the language in which individuals and society express them (Winch, 1958). It is argued that the reconciliation of generality and precision in the formulation of laws of human behaviour is an intractable problem. Economics, perhaps the most theoretically well developed of the social sciences, was long seen by some as having only a tenuous relationship to real behaviour except in a very few social systems, because it ignored the variability and the 'culture-relative' nature of human values (Lessnoff op.cit.). Accurate measurement and comparisons of behaviour are thought by many to be possible only in relation to cruder variables. However, economists have increasingly turned their attention to understanding the 'softer' determinants of behaviour such as esteem (Brennan and Pettit, 2004) and to the measurement of non-marketed goods such as health improvement (Williams, 1997).

As the systematic study of human behaviour has been extended, the limitations of 'hard' scientific criteria have become more strongly felt. Interpretative, illuminative, ecological and anthropological studies depending on internal logic rather than on external controls have intensified. The analysis of discourse and language has been widely taken up. Such methodologies raise basic questions about the nature and possibility of generalisation in social science.

However, it has been argued that there is room for some accommodation between this set of polarised positions, that it is possible to enter imaginatively into experience within a framework of some rigour, and that discipline of this kind can create productive and generalised theory of human behaviour (Brown and Harris, 1978). This may be the line of argument through which more fruitful co-operation might develop between the social and natural sciences in a period when the latter have abandoned claims to absolute knowledge and explanation. Learning disabilities, mental illness and human ageing readily come to mind as test areas for such collaboration between different networks or 'hybrid communities'. But the scientific issues remain largely unresolved.

One area of research which has provoked strong controversy about scientific method is that of evaluative studies. Anxiety about the effectiveness of evaluative research in policy programmes erupted amongst policy-makers and researchers in the United States following the huge investment in independent research appraisal of the poverty programme of the 1960s. Marris and Rein (1974) described the problems entailed in reconciling the aims and methods of researchers and programme directors. The precise clarification of aims, consistency of intervention and the time required for researchers to achieve rigour, objectivity and authoritativeness pulled against the indeterminacy and evolution of aims, flexibility of intervention and immediate feedback required by the programme participants. They saw rigour as in direct conflict with usefulness. Others ascribed the problems in part to an over-reliance upon quantitative methods and to an over-emphasis on outcome rather than process and context (Campbell, 1979).

Out of such problems emerged intervention studies or action research. These involved another set of boundary issues, between research and development. Does a project set up to evolve and evaluate alternative systems of care for the elderly constitute development rather than research? If so, it may fall outside central government's terms of reference and be the funding responsibility of local authorities. The boundary question is of practical importance. Mission-oriented researchers and policy-makers may agree that in some fields it is not further research but innovative development projects that are needed. But the agreement may be wider than this. Social scientists who question the possibility of creating significant generalisations about cause, effect and intervention in human behaviour may also question the validity of maintaining a firm separation between research and development.

Closely related problems are those of the relationship between research and dissemination and research and impact. They entail different levels of argument. There is, first, the question of resource allocation between research and dissemination. A leading social science researcher indicated to us that a field in which he was asked by the DHSS to consider further research had already had about £3 million from several sources spent on it and ignored (IE (80) 3). Could yet more be justified without first tackling questions about dissemination and impact? Some researchers insist upon time and money for dissemination of the results of their work; others build dissemination into the research itself. Boundaries between research, dissemination and service become blurred. Alternatively, emphasis may be laid upon the need for researchers to test and disseminate their work through participation in professional education.

The underlying issue is the nature of the impact of research. The 'stimulus response' model of impact (Nisbet and Broadfoot, 1980) and rationalist models of decision-making are largely discredited. Validity and reliability in investigation are not considered enough to change behaviour. A considerable body of work has been built up during the 20 years since this study on understanding and extending the impacts of research (see, e.g. Gibbons and Georghiou, 1987; Buxton and Hanney, 1996; van der Meulen and Rip, 2000; Molas-Gallart et al., 2000; Hanney et al., 2003). A particular focus has been the concept of impact itself and the development of alternatives to 'direct use' of research. One approach has been to reconceptualise impact in terms of processes, 'diffusion channels', 'routes towards impact' (van der Meulen and Rip, ibid.) and more generally forms of knowledge transfer (Lavis et al., 2002; Lomas, 2000). Here recent work on impact draws on earlier identification of the importance of personal relations in research use (Gibbons and Johnston, 1974; Platt, 1987; Pavitt, 1991) and on the burgeoning use of the concept of network in the science and technology policy field, primarily as developed in sociology and economics (Coombs et al., 1996). Brokerage, a concept developed in our 1983 edition, was also subsequently found to have significant influence in the role of research in policy-making (see, e.g. Henkel, 1994).

Some work was directed towards modes of disciplined inquiry that persuade policy-makers and practitioners to see their world through new eyes and to engage their commitment and action on that base.

Weiss (1977) offered a range of models of social research impact that are alternatives to direct, linear or instrumental models. They include the *political model* in which research findings become ammunition for particular interest groups and the *enlightenment model*, where social science research permeates the policy-making process not by specific projects but by its 'generalisations and orientations percolating through informed publics', so shaping the way in which people think about social issues. There is then the *interactive model* in which social scientists enter the arena of policy development as one of a number of participant groups. The process is then 'not one of linear order from research to decision but a disorderly set of interconnections.' Such models, though they refer specifically to social science, reinforce the notions that boundaries between science and society are permeable and that the power and authority of science may be played out in various ways.

Similar themes emerged in the literature on science policies in the 1970s. The pursuit of objectives such as the arrest of environmental pollution and inner city decay was perceived as requiring understanding and control of social, political and economic as well as physical forces. Salomon (1977) spoke of 'mobile targets' which cannot be 'hit' once and for all like the delivery of a man on the moon. Trist (1972) identified domain-based research in an era of turbulence as requiring new kinds of research institutions that are elements of learning systems, in which feedback and interaction between themselves and government can enable government to assume an anticipatory as well as corrective function. Problems are not likely to be 'solved'; rather they are anticipated, defined, elaborated and engaged. Traditional distinctions between means and ends become redundant. Bell's thesis on the nature and source of innovation in post-industrial society emphasised its dependence upon theory (Bell, 1973). Again, boundaries between theory and application, means and ends, science and technology and between natural and social science are all blurred in these conceptions of science policy.

Conclusion

This chapter has highlighted challenges to internalist theories of science as a unified, secretive, self-regulated, unequivocally authoritative system of thought and as a social organisation impermeable by external norms and influences. Science is depicted as heterogeneous in its modes and stages of development, in the relationships between the cognitive and organisational structures of its disciplines, and in the norms and motivations by which its practitioners are influenced. The boundaries between science and other institutions become blurred as externalist theories of science are accepted as having some force. Within this context, the question whether science can be steered cannot be given a yes or no answer. The stage of development of the science is a critical variable. Indeed, as sharp distinctions between fundamental and applied research are challenged and positivist theories of social science questioned, the concept of steerage itself begins to look too simple. A variety of potential relationships between science and research and government emerge.

Traditionally, British scientific policy was based upon internalist views of science and a hierarchy of status between fundamental and applied research. As government became more interventionist, it looked to science for salvation. As it then became more suspicious of the contribution that science could make to the problems of society, it set limits on the resources offered to fundamental research, and sought to 'steer' more applied research. The experience of so doing exposed the persistence of traditional assumptions and the power entrenched behind them. Our story of the DHSS's research commissioning following the Rothschild Report underlines the heterogeneity of the science which it attempted to harness, and the complexities of its relationships with policy. In some parts of the system efforts were made to grapple with them. But, as the complexities displayed themselves, there was a retreat to internalist theories or to one-dimensional and instrumental concepts of relevance, married to traditional patterns of power and status within science and between government and external institutions. The interconnections between epistemologies and institutional relationships constitute a recurrent theme of our story.

Chapter 3: Theories and Practice of Government

BRITISH PUBLIC administration long rejoiced in a style of relaxed pragmatism. But growing demands upon government entailed a shift from the tradition of intuition. During the 1960s and 1970s government faced conflicting forces. It searched for rational techniques in a period of growing social and political turbulence and of increased demands for participation by, and interaction between, many more extra-governmental groups in the processes of governing. Aspirations of rationality and participation themselves influenced changing fashions in policy planning. Many of these strands emerged and changed direction, if in esoteric forms, in the story of the DHSS's commissioning of disciplined inquiry. We consider this context before sketching some of the theories of government which throw light on the functioning of the DHSS as described in the next chapter.

The tradition of intuition

Government's relaxed and disinterested sponsoring of science which lasted until the early 1970s, through the University Grants Committee, through the research councils and through government direct commissioning, was a product of a system that was itself more relaxed and disinterested because confident that science and scholarship would do best if left to their own devices. This view was, for the most part, congruent with the dominant perception of the role of government. Government, before the second world war, did not seek to discover the *summum bonum,* or search for the good life on behalf of the whole society. The hidden hand could do most of that. Government was detached from the principal conflicting forces that affected the economy, the fabric of social structure and the services meant to alleviate its problems. Civil servants advised and acted from an intuitive sense of what seemed right, both logically and morally. Uncluttered by technology and its accompanying jargon, and within the generalist tradition, administrators were educated, rational, equitable, collegial and self-confident. 'Detached, at times almost aloof, he [the civil servant] must be if he is to maintain a proper impartiality between the many claims and interests that will be urged upon him.' (Bridges, 1950).

Government was primarily concerned to make the rules, to hold the ring (Kogan, 1971) and to maintain the essential fabric of society through the police, through grants meeting part of the costs of local government services, and through the defence of the realm. Issues demanding a definite policy commitment, as did the development of Concorde, the constant pondering over the best organisation for a National Health Service, were largely in the future.

With the growth of tasks placed on government as it moved on the road from a regulatory to an interventionist state in the second world war, and in the aftermath of the social reforms inaugurated in the 1940s, intuitive judgement began to be supplemented, though certainly not supplanted, by a style based more on rationality. Government began to do more and felt the need to know more, or at least to justify the knowledge it had.

Social and political turbulence

If, as late as the 1950s, certain assumptions about the ability and legitimacy of government to govern, and assumptions about what were reasonable resources for those purposes were taken for granted, in the 1970s not only did different social forces conflict more sharply than hitherto, but the very ground over which they fought shifted. The rules by which conflict might be reduced became uncertain. From every side increased demands were made on government, at all levels, to be more open and responsive to external forces, and as the rules changed so did the knowledge needed for their administration. Intervening directly and setting the rules were now intractably difficult and almost always contested.

Participation and interaction

That turbulence was part of the social scene was recognised by civil servants called on in the 1970s to deal with the pressure created by different interest groups (Nodder, 1979; Pile, 1976). The messages pressed on the centre were that the exercise of power should be through persuasion; that decision-taking was a result of interaction with widely disparate forces stating irreconcilable demands, a phenomenon clearly noticeable in, for example, the National Health Service (NHS) (Kogan et al.,1978) and that rationality was no longer enough.

Government moved from an era of working within implicit consensus towards a period in which the demands were for pluralism, for multivalue inputs to policy-making, for decisions to be carefully negotiated if they were to hold (Richardson and Jordan, 1979). Its ability to impose coherent social decision-making began to be eroded. Participation was allowed to groups not involved in managing the National Health Service through the creation of the community health councils. The creation of the Health Service and Local Government Commissioners in the 1970s introduced other forms of external review. There was the enormous growth in power of the public service unions, representing employees in services within the domain of the DHSS. And, at the same time, as both rationality and moves towards greater participation reached the centre, often contradictory theories of accountability were imposed, particularly on the Health Service. Some of us, in a study of policy-making in the NHS for a Royal Commission (Kogan et al., 1978) described this as a 'values overload'.

Planning theory and practice

There were also different phases of central planning philosophy and practice. Within the Health Service, as in other fields such as education, planning came to be seen as being as much a political as a technocratic and rationalistic procedure. Governments moved towards systematic indicative or 'first generation' planning in which demographic factors and potential resources could be placed against judgements on objectives and the best ways of achieving them. This 'first generation' planning was essentially quantitative and technocratic. It was synoptic in ambition. Synoptic planning assumes that it is possible to compare units of welfare, that it is feasible to take stock of what one is doing and to scan a large range of options in such a way as to predict how human beings will react to new and carefully planned systems. But uncertainties of purpose and method were later joined by increasing doubts about the ability of governments to manage economic futures. There were the movements in planning philosophy that might be generally called 'indeterminacy'. In the writings of Charles E. Lindblom (Braybrooke and Lindblom, 1963), Robert Dahl (Dahl, 1971) and Aaron Wildavsky (Wildavsky, 1966, 1979) attacks were developed against synoptic planning. The indeterminists instead preferred what Lindblom and Braybrooke called 'muddling through' and 'disjointed incrementalism'; in justification of their arguments the sceptics could point to the impacts of high-rise blocks of flats, motorway planning systems, large educational units and hospitals. The work of Lindblom, Dahl and Wildavsky also contributed to the questioning of processes. There is conflict between technocratic and participatory or 'polycentralist' planning which is concerned to determine what members of the public want rather than what seems 'logically' right and possible within the frameworks of administrators. There were changes from forecasting to Foresight and the elaboration of multiple futures.

The alternative or 'second generation' mode of planning, developed in parallel with first generation planning rather than sequentially, responds to multivalues analysis.

Options and uncertainties are fully displayed. Decision-makers have a wide range of perspectives open to them and can respond more intuitively to pressures from a more openly participative system. Planning is from, and for, uncertainty and ties up with participative systems which are polycentral rather than unitary.

In principle, the move towards recognition of uncertainty and the opening up of processes to accommodate different modes of thinking should cause government to leave space for other groups to participate more actively in the discussion of policies. But there is more than one response to uncertainty. It is possible to seek relief by recruiting help from outsiders. Or it is possible obstinately to dig in and stick to those certainties that are manageable. During most of the period of our story the first course of action was adopted by the DHSS. Yet while it was open to scientific help, the strengthening of the DHSS's planning capacity hardly seemed to relate to its research commissioning. Connections between policy and research based on policy analysis were not made. And moods of certainty and ways of dealing with uncertainty were cyclical. Moreover, as complex problems such as the crises of the inner cities and the competing demands in the Health Service for technological development and for community medicine emerged they displayed no evident logical, statutory or administrative boundaries; they belonged to domains within which several logics, value sets, technologies and power groups were relevant. In these circumstances the policy system became prey to the possibility of mutual vetoing or 'non-decision making' (Bachrach and Baratz, 1970) through actively preventing issues from reaching an agenda or crowding them out. The system was consistently under strain as it tried to collate, allocate and at the same time remain receptive. One way in which it attempted to deal with the cross-cutting demands of undisciplined politics or pressure groups and new patterns of social behaviour was through the development of coordinating or brokerage roles between government and the complex world outside.

The growth of rationality

The growth of rationality in British government has often been described (Garrett, 1972; Keeling, 1972; Booth,1979) and needs only brief summary. The first attempts derived not from abstract needs to predict the future but from dissatisfaction with somewhat incremental ways of controlling public expenditure. The procession of changes began with demands from the outside for change in the Treasury. In 1961, the *Plowden Committee on the Control of Public Expenditure* (1961) recommended the creation of the Public Expenditure Survey Committee (PESC). In 1957-8 a Select

Committee (Select Committee on Estimates, 1957-58) had noted 'that an obsession with annual expenditure can stultify forward thinking.' So, from 1961 onwards, forward looks prepared in annual cycles, but extending five years ahead, and over the whole field of public expenditure, were to take place. The underlying assumptions were that: 'Public expenditure decisions ... should never be taken without consideration of (a) what the country can afford over a period of years having regard to prospective resources and (b) the relative importance of one kind of expenditure as against another.' (Plowden op.cit). The Committee called for improvements in the tools for handling public expenditure problems and recommended the more widespread use of quantitative methods. Plowden recommended that ministers should be enabled to take collective decisions more effectively on public expenditure.

The next stage involved the acceptance from across the Atlantic of the whole bundle of rationalistic devices known as planning-programming-budgeting-systems (PPBS). These were first applied in 1961 to the US Department of Defense and then in 1965 to all federal agencies. Essentially, the components were the statement of objectives that had hitherto been implicit, the reduction of objectives to specific programmes, projected or actual, the analysis of the objectives to which existing programmes were directed, and the placing of programmes within a framework of planning and forward expenditure budgeting. There remained the problem of meeting the Plowden prescription that the usefulness of expenditures should be compared with their costs. The principal notion then employed was that of cost-benefit analysis. This was intended to provide to public sector policy-makers what they had hitherto lacked: information on the value provided to recipients of services in a system of costs and benefits where there is no feedback through the market.

Within the DHSS, programme budgeting began with a feasibility study during 1971. It was related to client groups which also underpinned the RLG structure. The use of cost-benefit analysis (CBA) would, it was hoped, break away from the position in which

> decisions on the use of resources are based on costs but not on benefits . . . and organised around resources ("inputs"). CBA attempted to put a money value on unmarketed goods and services, including externalities, by studying choices which people make, or say they would make, which give some guide to what they would pay if there were a market. (Banks, 1979)

But, as Banks well relates, large public services are not as susceptible to analysis of individual choices as are other forms of spending and analysts must classify services into expenditure 'programmes', according to the benefits provided although wherever possible benefit is measured against the cost of the particular service. Even where benefits are not comparable, politicians might be able to put some kind of valuation on the choice of one thing as against another through a 'trade-off'. As we shall see, the DHSS scientific advisers attempted to infuse CBA calculations into some research projects.

The Heath government, as part of its attempts to improve and reorganise central government (White Paper, 1970) selected among the various PPBS approaches by introducing programme analysis review (PAR) which would provide studies in depth of both existing and alternative policies and programmes. It was hoped, then, that government could predict and work towards a comprehensive grasp of its main policies. The creation of the Central Policy Review Staff reflected expectations that the many different concerns of government could be synthesised so that decisions could be made within a pattern of healthy coordination. The Rothschild Report (Rothschild, 1971), too, applied assumptions handed down from a peak of confidence of government's ability to define policies.

The move towards greater rationality had other effects. Departments, sometimes reluctantly, shaped up some of their existing forms of analysis into separate planning units. Following the Fulton Report on the Civil Service (Fulton, 1968), attempts were made, many would say half-heartedly, to admit scientists and other non-generalists more definitely into the policy-making forum. How far science and non-administrative expertise gained a foothold in Whitehall must be a matter of doubt (Gummett, 1980). Certainly, however, the appointment of Chief Scientists, the bringing to parity of some specialist groups with their administrative counterparts, the attempts to commission more policy-related research than at the beginning of the 1960s, were all evidence of the intention to follow the spirit of Fulton. The institutional obstacles to such a rapprochement, however, proved strong.

We should not exaggerate the onset of rationality. White Papers sanctioned by a prime minister, such as those of Edward Heath in 1970, have dramatic effects and are signalled clearly in histories of any period. The response of a large and sprawling organisation such as the DHSS to the many pressures upon it is far less easy to document. In particular, other and conflicting contra-rational movements had their effects. At the same time as the DHSS seemed to make its contributions to policy ever stronger, even when, after 1979, a Conservative government declared its intention to 'disengage' from the field, there were the attempts to give more power to the periphery and growing yieldings of ground, as well, to different forms of participation.

Differentiation within the government machine

Government contains a spectrum of policy- and decision-making which made it difficult to achieve its ambition of 'academic research strategy' (DHSS,1979). At one end there is the function of setting values, performed mainly by politicians and generalists. At the other end, there are determinations based on technical and factual knowledge. The values-technical spectrum (Subramanian, 1963) assumes that government takes account of political and interest group pressures, the values of client groups and their professional helpers, at the same time as it absorbs technical facts. Technical facts constitute a wide range of their own. They might include the potential of a radiological scanner, or 'the burden of illness', or knowing whether a financial approval is likely to run foul of the auditors. The mediation between value and technology is held to justify the generalist administrator. It assumes that issues that require the sorting out and weighting of complex values take the administrator outside the areas of specialist and into generalist knowledge and skills.

But there is more than one kind of specialisation. Some specialists must also be generalists. There is specialisation by skills and by organisation. Organisational specialisation rests upon Weberian assumptions about the need to allocate responsibility in a logical and hierarchical structure (Burns and Stalker,1966). It has been assumed that by contrast, 'flexible and informal methods of team work ... rest more upon the skills and status of the various participants.' But it is by no means certain that a Chief Scientist or Chief Medical Officer will act less hierarchically than an administrative officer incharge of science or that team work is more a feature of professional work than of generalist bureaucracy.

An organisation divided by skills has problems of coordination. Thus within a government department accountability is focused to the minister through a single congruent point at the top, the Permanent Secretary. Moreover, a specialist who controls and coordinates becomes 'a generalised specialist'. 'His skill must. . . lie . . . in evaluating other specialist contributions ... within a framework of certain accepted techniques and objectives.' He, as much as any lay administrator, is subject to the administrative and political process (Self, 1972). Those who have authority to collate advice and information from *anywhere* along the spectrum of values and technology take on the role of appraiser, or evaluator, and act as broker and collator of considerations distant from their own expertise. The skills of the appraiser (ibid) are deemed to include political realism, the ability to simplify and make decision-making tidy, to clarify duties and responsibilities, to work within precedents and expenditure controls.

The policy-making process

The most general perception of government and of the civil service is that of stern, efficient and rational convergence. In this vision, government is depicted as the authorities who make policy which is defined as the authoritative allocation of values. Government does this through regulating a flow of inputs which are essentially demands made upon the allocators by the wider society. Decision-makers thus receive demands from outside themselves. They reduce interests, they act as gatekeepers who refine, or allow, or obstruct the traffic of demands up and down a system.

This view fits well the traditional view of the British civil service which enables, limits and formalises by presenting constraints to ministers who wish to introduce or implement policies, and marshals into logical and resource frameworks the demands put up by external pressure groups.

The systems metaphor is useful because it directs attention to the way in which values are located and held, and how policy-making, and actions resulting from them, require interests to be heard, and then reduced to manageable proportions. Its main fault is that it assumes government itself to be stable and rational (Jones, 1966) a system in which interests have a hearing, and authoritative actions result. Although it provides for feedback mechanisms through which the results of policy implementation are able to affect the next allocation of values, it need take no account of the internal dynamics of government itself. It assumes the mechanism to be neutral so that values setting within, and the dynamics of different groups struggling to make their viewpoints heard, are discounted.

Alternative descriptions

Alternative descriptions of government contest the assumption of a neutral system by emphasising the elite, self-regarding and internally-motivated nature of those who maintain the system. So, far from serving ministers whose values predominate, or enabling values to be identified, aggregated through interest reduction and then allocated and adjudicated, the permanent instruments of government become an interest group who might indeed co-opt others into loosely-coupled coalitions of elites. Government is primarily concerned with the maintenance of power of its practitioners or even of the class which they are alleged to represent.

There are other models implying deviance from neutrality. Some point to the recurrent tension between the politicians who are accountable for government actions and civil servants who present obstacles to the political will. Officials have their own biases or their own modes of analysis, or their own needs for long-term negotiation and compromise, which make them see matters differently from the ephemeral and heroic minister. Or, again, central government's neutrality is thought to be subordinated to Treasury norms of financial conservatism so that it is not able to take a properly reflective view of the 'real' needs of the clients and of the services over which it has differing degrees of control (Kogan, 1969).

Our own view is that all of these descriptions have something in them. The systems view of government is an ideal model applicable to a stable democracy which seeks to agree on ways to get things done more than on what should be done. Politicians and their administrators are indeed supposed to elicit support, to cope with the range of values presented to them before reducing interests, and to ensure that the outputs match the intensity of the inputs and are then properly distributed and adjudicated. That is, roughly, what western democracies expect of their administrators. It is also true, in real life, that if society is polycentral, pluralistic, and just confused, and that if no man is an automaton without feelings or personal values, then the bureaucracy is most unlikely to be able to suppress its own pluralism and the values that it generates through the performance of particular tasks. Our account of the DHSS's relationship with science largely confirms a modified view of pluralism in which such scientific elites as came onto the scene were at strongest 'loose-coupled' with the policy-makers.

If the DHSS, or other central government departments, have their own pluralism their own internal negotiative patterns which they have to resolve, it becomes all the more difficult for them to act as an efficient but responsive 'receptor' of the pluralism of the world outside. Hence its difficulties in presenting itself to the scientific community in order to ask for disciplined inquiry that might contribute to policy-making. Hence, too, its difficulty in focusing on all of the pluralisms that are represented by science. Its RLGs could work; its CSRC could not, for these reasons. In acting as a receptor and as a system stimulating other prime agents the Department cannot easily act convergently.

Before considering how various factors analysed in this chapter provided the context for our empirical study of the Rothschild Experiment in the DHSS, it is useful to assess how some of the key trends identified continued to impinge on the organisation of government in the period since 1983. At an analytical level, there was a continuation of work on the nature of institutions. This provides several potential frames for denoting the origins and roles of institutions. Theories of institutionalism are contended between academic traditions; between historical institutionalism, rational choice institutionalism, and sociological institutionalism (Hall and Taylor, 1996). In depicting how policies are shaped by the nature of institutions and howinstitutions generate norms such theories, if not directly applied to particular organisations, throw light on the structures analysed in this volume.

As noted, there were changes after 1979 and part of this reflected the increased influence of public choice theory, at least among economists and political theorists. It involves the application of the economist's way of thinking to politics. It could be inferred that many of the government-led reforms over recent decades have been inspired by this kind of analysis. In what academics identified as New Public Management (Pollitt, 1993, 1995), eight elements comprised a shopping basket for those who wish to modernise the public sectors of western industrialised societies (Pollitt et al., 1997): cost cutting, capping budgets and seeking greater transparency in resource allocation (including formula-based funding); disaggregating traditional bureaucratic organisations into separate agencies; decentralising management authority within public agencies; separating the purchaser and provider functions; introduction of market and quasi-market type mechanisms; requiring staff to work to performance targets, indicators and output objectives; shifting the basis of public employment from permanency and standard national pay and conditions towards term contracts, performance related pay; and emphasis on service 'quality', standard setting and 'customer responsiveness'. Similar lines of thought were advanced in the Thatcher government's adoption of Hayek's thinking and its acceptance of the arguments for the marketisation of public services (Kogan and Hanney, 2000).

These notions were used in different settings in a pick and mix fashion, and were not presented as a single and identifiable bundle, but New Public Management was relevant to the organisation of publicly-funded research as it moved towards transparency in resource allocations, standard setting, responsiveness to consumer wishes, different forms of centralisation and decentralisation in different directorates. We return to its implications for the organisation of NHS R&D, and of the NHS itself, in Chapter 13.

Related developments that are particularly relevant for the organisation of research, as well as government more widely, include an even greater emphasis on user perspectives, often at the expense of producers and their interest groups. The user perspective has been expanded to include the public, but this is variously interpreted to include the general public, or, more narrowly, individual patients and/or their interest groups. Several modes of citizen participation have long been proposed (Arnstein, 1969), but they are now of even greater relevance to this debate because the focus of participation is no longer just on health care policy making, but has spread to decision-making about health research. Arnstein's ladder of participation has been adapted in accounts of the recent development of public engagement with health research (Pleasant et al., 2003; Oliver et al., 2004). For policy-making in controversial but highly technical fields it is increasingly seen as relevant to engage the public with the science and develop mechanisms such as citizens' juries to gather their opinion (Weale, 2001; Irwin, 2000). Furthermore, and as also noted by Weale, the increasing role of the European Union (EU) is adding yet further layers of consultations and competency in various areas of public policy-making. This has gone further in research policy than in health care policy, but multi-level, multi-actor approaches can be applied when analysing the increased complexity (Kuhlmann, 2001; Marin and Mayntz, 1991). A further development, especially since 1997, has been the attempts to evolve evidence-based policy (Davies et al., 2000). Illustrating the multiplicity of pressures within the system, the move towards evidence-based policy represents something of an attempt once again to strengthen rational approaches to policy-making.

Processes

Returning to the policy processes as they were seen at the time of our empirical studies in the 1970s, we have already noted the critique of government as rational, represented particularly by Lindblom, Wildavsky and others. To their minds, decision-making is incremental, disjointed, episodic and incapable of making a synopsis of needs and wants so as to produce rational planning in which ends are well adjusted to means. Instead, policy-makers have to satisfice (Simon, 1945), external pressures are taken into account and policies follow reasonable and minimal feasibility rather than a vision of the 'good life'.

The schematic metaphor of the input-output model, advanced by Easton (1965) and others, does not of itself describe policy process. Various metaphors have been adopted. Schon (1971) raises the question of how norms set in the external environment enter into the internal system. He is particularly concerned with the normative issue of how the policy systems can become learning systems and thus ensure that there is a movement from the periphery to the centre, that a procession of ideas generated outside the decision-making structure can enter the arena where decisions are made and take a legitimate place in them.

Other notions of the policy process (eg Hill, 1979; Richardson and Jordan, 1979; Dunsire, 1981) emphasise that there is no clear end and no clear beginning. The process itself is virtually indistinguishable from the product (Rein, 1983). The ways in which power is exercised, and for whom, are as important as the purposes, overt or implicit, for which power is exercised. This type of thinking goes beyond the systems theorists' use of the notion of feedback because it allows for the process to be emancipated from the notion of a system which, is self-correcting, equilibrium-seeking and capable of consensus. The interaction between policy-makers and the environment within which they work has been summarised as being either situationist or individualist or interactionist (Linder, 1981) The situationist emphasises the extent to which environment presses upon policy-makers so that, as in budgetary processes, the constraints presented by the environment predominate. The individualist version allows a somewhat more heroic role to the policy-maker who battles his way through the established norms and rewards - a view particularly adopted by those who look to the economics of the small firm as a model of bureaucratic behaviour. The interactionist - surely the more realistic model? - notes how the environment imposes constraints upon the individual and how individuals themselves constitute the environment for at least other individuals and affect the rules and the environment in which they work. The interactionist position is well illustrated in our view by some of the dramatic changes in structure, style and outcome which are attributable to individual action moving against a greatly changed environment in the story of the DHSS research commissioning system.

It would be possible to use any or many of the heuristic simplifications briefly related above. In practice, the analyst must shop around among current explanations to see what language or concept best fits the case to be described. Our own perspectives owe a lot to systems theory, but also to current views about politics as a somewhat inchoate contest leading at best to negotiated orders, and to decision-making that most often negates the rationalist hope of defining ends and achieving them systematically. There are institutions and elites, and systems but they are loosely coupled through the media of exchange, dependency and negotiation.

Essentially we consider central government to display two conflicting characteristics. The prevailing stated norm is that of convergence, unity and equity within a system that must act authoritatively. But the behavioural environments impinge upon a system whose essential pluralism reveals itself, perhaps, not only through variation at any one point in time but also through volatility over time. Thus a single policy might take hold authoritatively but just as quickly give way to another policy.

The constraints laid down by ministers, parliament, the economy as mediated through the Treasury, and the environment of public opinion are boundaries within which administrative discretion is employed. But they are changing and permeable boundaries. At the same time individuals may occupy a heroic role. Certainly the actions of ministers and of the three Chief Scientists during our period have critically affected the way in which policies have been determined.

Our propositions are, then, of several kinds. Different groups within both government and the science community are influenced by different value positions, different knowledge systems, different power relationships and different phenomenologies.

In order for plans to be made, programmes to be achieved, policies to be stated, some reduction and convergence have to be achieved. This happens through different forms of transaction. These include negotiations in which both sides have something to exchange so that *quid pro quos* can be made. A degree of mutual dependency develops. But the conditions under which free exchange and mutual dependency occur changed. Previously, a science system had freeholds from which to negotiate. Growing uncertainty in higher education meant that even those with tenure could not be sure what resources they could deploy without reference to government. There were coercive relationships in which unilateral authority or superior bargaining power was used. It was not therefore a true political system in which gaps between expectations and resources could be negotiated or fought over. The power was too much on one side for that.

Much of what happened could be explained by the imagery of exchange theory which shows that when different groups are uncertain of their share of resources they seek either to create, or to avoid, dependencies on other (Blau, 1964). Ranson describes how different groups seek to maximise their power and authority through exchange relationships: 'They may seek to possess strategic resources which others may desire; secondly, ensure that these resources are scarce and unavailable elsewhere: third, have the capacity to use coercive sanctions if necessary; and lastly, be indifferent to the resources possessed by other actors.' (Ranson, 1980). Where there is a positive imbalance in exchange, power accumulates on one side.

In our case, had the full Rothschild arrangements succeeded, the scientists would have been in a position to give something that the DHSS needed and could not easily get elsewhere. But the conditions under which exchange took place became weaker. Scientists became more dependent on directly allocated government funds. The scientists' currency was debased as government became increasingly indifferent to what scientists could put into the exchange. Scientists could not produce a contribution which was either indispensable or the result of a collective and authoritative effort over the whole range of science. Science was thus unable to be a full partner in the exchange, and could not generate the power of a partner with something valued to offer. Those parts of science without which government could not manage, such as those concerned with medical research, were able to hold their corner. But, as we shall see, the Chief Scientist's Research Committee, with its primary responsibilities for health and personal social services research, became unequal to the burdens placed on it.

With this background assessment of the nature of government and policy-making and its modes or approach to outside help we can now turn to examine the particular case of the DHSS.

Part II The Commissioning System in Action

In this part of the book, from Chapter 4 to Chapter 10, we describe how the DHSS attempted to recruit the help of science and the structures, roles and relationships that resulted between 1974 and the beginning of 1981, although we take the history back to its natural starting-point of 1972 where necessary. Each of these histories yields generalisations which have starting-points in the theories of science and government which are examined in the preceding chapters.

We first describe the DHSS as an organisation for making and implementing policy. Our analysis of its structure displays the ways in which it embodied the multiple values and policy initiatives emerging from the time in which it had to work (Chapter 4). In Chapter 5 we take up the first of our detailed cases: the creation, working and eventual abandonment of the Chief Scientist's Research Committee and two of the Boards beneath it. Here our dominant theme is the nature, and the limitations, of macro scientific policy. The case also enables us to scrutinise in some detail the nature of the exchanges made between policy-makers and scientists. This latter theme is, indeed, further exemplified throughout the rest of our empirical data of observed cases.

If the story of the Chief Scientist's Research Committee is that of uncertainties on both sides, the story of the DHSS's relationships with the research councils, as particularly exemplified in the bizarre story of the Panel on Medical Research (Chapter 6), demonstrates the contrast between the uncertainties of the policy-makers and the certainties of internally-governed and imperially-defended medical science. In sharp contrast to the stories of disappointed expectations is our account of the Research Liaison Groups (Chapter 7) and the Small Grants Committee, where the differing values, epistemologies and institutional structures of policy-makers and scientists did not, after much work and committed effort, inhibit a rapprochement and working relationship between both sides. This causes us to ask why policy-science relationships work at particular levels of problem-setting and solving and not at the more abstract levels.

In the final chapters of Part II we analyse the nature of the research units established from the 1960s on DHSS grants. In Chapter 8 we first illustrate the enormous range of publicly related science undertaken by the units. In Chapters 9 and 10 we explore the twin but sometimes competing concepts of scientific merit and policy relevance. These abstractions come into political and institutional interplay in these chapters as we recount our findings from our study of the Department's reviews of research units. In these chapters we make good our generalisation about the multiple criteria for science that can and should be set and the ways in which its institutionalisation can narrow the range of scientific acceptability. Different kinds of scientific activity generate different networks and different degrees of power and authority.

In these chapters, we show how the DHSS threshed and turned in its attempt to make a good match between science and policy, and we believe that both its successes and failures yield generalisations that are likely to apply to similar attempts in other domains of public policy.

Chapter 4: The DHSS and the Research Management System

AS WE have seen, in the 1960s and 1970s the DHSS had to operate in a milieu of increasing social and political turbulence at the same time as rationality and positivism came into vogue. Whilst most government departments felt the impact of changes in the larger society, the DHSS exemplifies more than most the procession of values which affected the machinery of government. So it has been successively restructured to reflect changing functions and processes, new preoccupations with client groups, rather than institutional frameworks, rationalist planning as well as the transactions with the field authorities.

The former Ministry of Health became the Department of Health and Social Security responsible for health services, for personal social services and social security. Underlying this amalgamation was the assumption that the health, income and personal social welfare needs of clients were strongly linked. Whether services should emerge as cash or kind, income support or publicly-produced services, or whether there could be a decisive shift from institutional to community care, were issues supposed to be better resolved through administrative proximity. Such linkages and resolutions were, however, somewhat tenuously achieved in the DHSS and then only through meetings at the strategic summit.

The same trends affected the NHS itself. The hospital service under its own regions, environmental and community health under local health authorities, and family practitioner services under their committees, were replaced in 1974 by a unified health service. It was the DHSS which both led and had been the recipient of the astonishing range of changes in its organisation, policy- and decision-making patterns. The DHSS encouraged the creation of elaborate planning systems at the same time as it required participation of health service workers in planning and incorporated the client voice into policy critique through community health councils.

The DHSS came to preside over the newly-combined children's mental health and welfare functions as, following the Seebohm Report, the local authority social services departments were created in 1970.

The Department then struggled to get co-operation between health and social services authorities. Joint planning as well as overlap of memberships were promoted.

In all, the Department therefore acted as host to the impressive range of political values which emerged in the 1960s and 1970s in Britain. At one and the same time, it attempted to promote accountability, acceptability, participation, delegation, efficiency, coherence, comprehensibility, certainty, rationality and feasibility (Kogan et al., 1978).

The main groups of civil servants with whom our story is concerned were the administrators and the professionals. The administrators were the element common to all parts of the DHSS. They worked within civil service norms of hierarchy, and parliamentary and financial accountability. They responded simultaneously to the policies of the wider government network led by the Treasury and the Cabinet Office and to the demands generated by the DHSS field authorities and pressure groups. They had to collate policies developed with professionals and specialists and put them into their frameworks of law, finance and the service objectives established by ministers.

Policy outcomes are the product of both the mainstream administrative values of the civil service and political activity which is, however, 'like lightning in that it may suddenly strike into any corner of the administrative system but only rarely does so.' (Self, 1972).

Professional groups in the DHSS played an exceptionally important role, but their weight varied between different policy fields. Few professionals worked with the social security divisions. The most powerful were probably the medical officers under a Chief Medical Officer with Permanent Secretary status and four Deputy Chief Medical Officers ranking as Deputy Secretaries. The Chief Nursing Officer was of Under-Secretary status and the Chief Social Work Officer who headed the Social Work Services, a Deputy Secretary in rank. The OCS had the equivalent of 10.5 full-time professional officers in 1980, five doctors, four social scientists and 1.5 nurses.

Professionals within the DHSS had power rather than authority to decide; that rested in the hierarchical line between ministers and the administrators. To some extent their power depended on the technical content of the issue to be determined.

The dominance of administrators weakened the more esoteric the issue. Once a matter could be reduced to judgements of value comprehensible to ministers or administrators, the professionals' power became weaker. This may explain why professionals were less willing than administrators to subordinate the work of the research units to the needs of the customer divisions. They empathised more closely with the needs and rights of researchers to pursue their own norms and needs (ID (76)20, 26, 32); but their knowledge might compete with that of their fellow professionals in the field or with scientists who could bring to the Department alternative forms of knowledge or analysis.

In the DHSS the traditional patterns were restructured on the advice of the McKinsey firm of management consultants. The policy divisions represented a particular analytic cut-through of DHSS interests. The distribution of work was first systematically discussed in the Haldane Report (Haldane, 1918). Haldane thought there were two principles only upon which the Department's functions could be determined and allocated: one was the distribution according to the persons or classes to be dealt with, and the other the distribution according to the services to be performed. Haldane came down on the side of allocation according to services. This mode predominated until two potentially conflicting principles were introduced in the late 1960s and in 1970. The McKinsey reorganisation of the DHSS introduced the concept of 'client groups' which were, in effect, a reversion to the pre-Haldane principle of distribution according to persons.

McKinsey's structure assumed that pressure upon the policy-making system would come from groups such as the mentally ill, the elderly or lobbies representing children's needs, whose demands might be presented not only through the field authorities but also on a national basis in a way that would cut through the divisions of health and social services. The DHSS thus established the 'client groups' which were now held together in the Service Development Group. Regional liaison presents the alternative theme of geographical units. That client group assumption might have been thought radical in its day. It was later criticised because it seemed to tie policy to the demands of particular client groups and the institutions and professions which cared for them, whilst it was felt that policy should be freer to move towards new formulations of institutions or professionalisation.

Decisions made by the divisions were brought together because they shared a departmental budget, a single staffing and promotions system, a single system of political control under a team of ministers and the Permanent Secretaries and Deputy Secretaries. They all responded to Treasury prescriptions on finance and to the patterns of establishments and organisation controlled by the Civil Service

Department. These entered the DHSS through the finance divisions, the administration group and the personnel divisions.

Yet, unified as the Department was, a division led by an Under Secretary was remarkably free to generate policies and control its own blocks of work. Problems, both operational and of broad policy, were first tackled there and many were, in fact, disposed of before they reached the more senior levels, though senior officers working under ministerial direction eventually determined policy. The formulation of policies began with the Principals and Assistant Secretaries and their professional counterparts. An Under Secretary would generally reckon to make quite important decisions within the general policies which he has helped to create. The discretion of administrative and professional groupings within the policy framework made the persuading, wooing and capturing of customer interest in research a key development in the Rothschild story.

The range of roles

Government produces roles that take up the functions and styles associated with them. One source of variation is their nearness to the decision-making directly affecting the services. There were administrators in the regional liaison divisions who interpreted policies to the field authorities. They had to respond to requests for approvals from the field and at the same time provide material upon which parliamentary questions, debates and MPs' letters could be answered. They represented the Department as a decision-making and authoritative body. The Service Development Group might make decisions, but only if they affected a whole client group. A bit farther along the spectrum were the professionals in both regional liaison divisions who interpreted policies to the field authorities. They had to respond to requests for approvals from the field and at the same time provide material upon which parliamentary questions, debates and MPs' letters could be answered. They represented the Department which might be technocratic and authoritative, or speculative and open, or all of these at once. It could encourage critique or counter-analysis of the decisions being produced by the politically accountable machine. And there were inputs to the policy system from those who were licensed to put up critique - namely, those who were engaged to produce research or other forms of disciplined inquiry for the DHSS, usually from the outside.

Although we have stated these characteristics as if they were separate, in each role we can find elements of all of them. Thus the desk administrator working with field authorities was not impervious to the larger planning implications of specific decisions. Service development was not an abstract process, but derived from awareness of what happens in the field. Equally, the professionals could not cast off the authority that the DHSS bestowed upon them and, indeed, professionalism need not imply a less authoritative or less closed system than that of administration. Our point here is that the Department itself precipitated a range of roles and a range of styles, a point strongly substantiated by what we have to say about the customer function later. Within the range of roles and styles, disciplined inquiry is a small and quite fragile part. The theme of diversity is reinforced if we look at psychological and social relationships in government. Officials are socialised into the norms of the larger department. There is a constant movement of administrators and, to a lesser extent of professionals,* between divisions so that officers have no chance 'to go native'. Instead, they learn to identify factors common to diverse policy issues, and to respect a common repertoire of competences. The competences are not trivial and *in toto* represent considerable expertise; the ability to render technical issues into statements that MPs or ministers or the public can understand, the ability and courage to make judgements on disparate value issues that may be incapable of objective or scientific comparison. Professionals come late to the Department from different traditions and have to find their place in an institution whose norms may be in conflict with the training and backgrounds of their peers outside the civil service.

These cross-cutting structures of relationships and roles form the context of the DHSS commissioning system created to implement the proposals of the 1971 Rothschild Report, to which we turn in the next chapters.

* By 1979, hardly any of the principal actors concerned with the commissioning of research within research management were the same as those with whom we first worked when we began our project in 1974. But most of the leading researchers were still in place. Almost all of the heads of the customer divisions had also changed.

Chapter 5: Science and Macro Scientific Policy: the Case of the CSRC and the Intermediate Boards

THE CREATION of the Chief Scientist's Research Committee (CSRC) and its two intermediate boards, and their demolition six years later in 1978, are an important example of encounter between government and science. In order to convey their significance and style we shall describe the main issues which these bodies faced, their membership and how they were appointed, and the balances between different scientists and practitioners on them. We shall also consider how the scientists and the DHSS perceived the roles of the various members. The broader themes of the exchanges of power and the mutual dependencies between the different groups, what was being given, what was being taken and what was being expected by each side, will be taken up in our final chapter.

The Chief Scientist's committees*

In the DHSS commissioning system as it existed between 1973 and 1978 (Figure 1.1), *the Chief Scientist's Research Committee* (CSRC) (Figure 1.1 (i)) was to concern itself with all aspects of DHSS-funded research in health and personal social services. Two intermediate committees, the *Health Services Research Board* (HSRB) (Figure 1.1 (ii)) and the Per*sonal Social Services Research Group* (PSSRG) (Figure 1.1 (iii)) were expected to perform functions similar to those of the CSRC within their subject areas. The *Panel on Medical Research* (PMR) (Figure 1.1 (iv)) acted in parallel with these two committees and we discuss that and the *Small Grants Committee* in separate chapters.

Membership

The great majority of committee members were scientists. Practitioners and policy-makers occupied a relatively small number of places. The CSRC's members were chosen to provide a reasonable balance between the main academic areas concerned with DHSS-commissioned research. Members were to be appointed for three years with an option on either side to renew membership. Some members did not continue after the first period but the balance of disciplines remained roughly the same through the five years of the committee's life.

* The whole of the system associated with the Chief Scientist was known as the Chief Scientist's Organisation (CSO). This should be distinguished from the OCS (Office of the Chief Scientist) which succeeded the Research Management Division.

When disbanded in 1978 the committee included five members from the classic medical sciences, three from social and community medicine, four from sociology and social administration or cognate social sciences, one from nursing studies, an economist and a social anthropologist. There was a Director of Social Services and a Regional Medical Officer who was replaced by a general practitioner. The Secretary to the MRC and the Chairman of the SSRC were *ex-officio* members. There were assessors from the DHSS and the SHHD.

The medically-based disciplines formed a majority but it cannot be said that they dominated the proceedings. A social scientist tended to take the Chair when the Chief Scientist was away. An economist member had a particularly strong influence over the committee's thinking. There was never a display of the in-built power of medical scientists or conflict over which projects, units or methods should be promoted perhaps because the committee never reached that point in its work.

How did the CSRC see its role?

In one of its earlier minutes, the CSRC stated an ambitious perception of its role:

> Apart from the task of establishing priorities between RLGs... the Committee reaffirmed its awareness of its responsibility ... for oversight of the RLG structure and for accountability to the Planning Committee for discharge of this stewardship, and its awareness also of the need to provoke research thinking in subjects which fall across a number of RLGs. (CSRC (M) (75)/4.8)

Members soon discovered how difficult it was to establish priorities for either science or policy concems on other than political and social criteria, which were not for them to determine. Nor could they easily discharge any kind of stewardship to the Departmental Planning Committee, consisting of the Permanent Secretary and other official heads of the office, which seemed hardly concerned about the CSRC's existence, let alone its work, and which afforded no channels through which communications could pass. As often, the outsiders on the CSRC assumed that government was more omniscient and systematic than it was. Its attempts to provide research thinking across boundaries made promising beginnings but time ran out before it could produce results.

What did the CSRC do?

The CSRC tried to elucidate policy priorities for research, to identify dimensions of disciplined inquiry which concerned more than one Research Liaison Group, to create controls and guidance over the RLGs and to establish appropriate methodologies. In the end, its most important efforts failed. The relationship between two highly sophisticated and elaborate systems, a government department and the world of science, was too difficult to achieve. Government and science could come together to tackle discrete and individual problems but not the arrangements for the government of the whole range of DHSS's science concerns.

These difficulties were reflected in the fact that the CSRC devoted a large proportion of its agenda time to discussing the organisation of research in the DHSS, the future work of the CSRC, its relations with RLGs, how to decide priorities and budgets, its relationship with the MRC and biomedical research, and, indeed, the tenure of CSRC members (CSRC (M) (74)/12).

Such issues occupied both its first and its last meeting (CSRC (M) (78)/4). The fact that it put so much effort into considering them confirms the intrinsic difficulty of getting right the relationship with the DHSS.

The CSRC began by endorsing, without substantive change, proposals put to it by DHSS administrators. It approved the setting up of RLGs and received reports on their working; it approved the establishment of a Small Grants Committee and arrangements for the management of funds transferred from the MRC. It heard reports on the specialised research programmes - building, social security, computers and supplies.

A great deal of effort was spent on its relationships with the RLGs. It also made several forays into broader questions concerning the role of science in policy-making. It was concerned to find both criteria and policy concerns that would cut across the demands voiced by the client groups represented in the RLGs. It declared at one of its earlier meetings that the research effort should be concentrated on the promotion of health and the prevention of illness (CSRC (M) (74)/7.6). It was at pains to make sure that such terms as 'health' and 'illness' were used in their widest context to incorporate social, medical and political factors.

The best utilisation of resources was an example of 'criteria' and terminal care was an example of a policy issue that touched the concerns of many RLGs (CSRC (M) (74)/8.13). Later in its life, the CSRC took on board a study undertaken by the Chief Scientist and one of the Department's economic advisers on the burden of illness

(Black and Pole, 1975). This led to the declaration that 'the health departments hold that, other things being equal, priority should be given to the common conditions and that research expenditure should be in rough proportion to their relative burden on the community.' (CSRC (77)/18). The Department thus endorsed a simple correspondence theory: disciplined inquiry would be best encouraged where the service problems were greatest. The committee did not dispute this. It took on other system wide concerns when it tried indecisively to formulate policies for research training (CSRC (M) (75)/12.9).

Some of its members saw it as their duty to review critically the work of the DHSS. One social science professor maintained that the Social Security Research Policy Committee could challenge the Department's uncertainties about the scope for research into social security. It might heal the 'structural and philosophical' separation of the two parts of the Department concerned with health and personal social services research and social security (CSRC (M) (74)/12.7).

Challenges to the Department on such substantive matters as the content of research programmes were a feature of earlier rather than the later meetings. A report of a building and engineering working party elicited discussion on whether design packages for building could be created, and whether there could be more research on the philosophy and practice of planning procedures to shorten the time of completing buildings. Work was needed, it was thought, on evaluating how effectively buildings served the purpose for which they were designed (CSRC (M) (74)/12.8). Scientists thus attempted to 'percolate' the predominantly works profession membership of a working party, not so much as scientists exercising technical expertise as intellectuals concerned with broadening the scope of the policy discussion. Their concerns included, too, the advancement of the cause of science in government. RLGs' membership should be increased to bring in the underrepresented disciplines, to enable more working scientists to become familiar with the research problems encountered by government, and to provide valuable experience for young scientists, a view, however, contested by existing RLG advisers when consulted by the CSRC (CSRC (M) (75)/7.9). Thus the scientists were concerned both with the problems set by customer divisions and ways of best bringing science to bear on them, and with inherently scientific concerns. They wanted to apply critique to government and to percolate its policy-making with their disciplinary and social point of view. They hoped to advance service causes.

Membership of the CSRC elicited various role constructs, some not contemplated by the Department when it recruited them.

Attempts to guide the RLG system

The CSRC decided early in its life in 1975 that it should give guidance to the RLGs. They were the main site of action (see Chapter 1). It would be valid, the committee thought, to suggest that RLGs should concentrate on research into alternative patterns of care which would advance the best use of existing resources. From then onwards, the CSRC worked hard, several times and in several ways, to establish frameworks within which it could exercise its strategic role over the RLGs. In these efforts, the committee's main preoccupations were with policy rather than with research content or methodology.

One member prepared examples of questions for the CSRC to put to the RLGs (CSRC (M) (75)/7). They were: should we concentrate more on alternatives to hospitals; on self-treatment; on substituting paramedical for medical personnel; on projects which would have both quick pay-off and low implementation costs if successful? RLGs should attempt, it was thought, to identify the major problems facing them in the next three to five years, and the information needed to come to grips with them. They could then identify 'the information gaps which could feasibly be filled by research.' The CSRC was willing to advise the RLGs that resource implications must affect which research to encourage. No simple pursuit of the scientifically interesting sufficed as a criterion. In the same year, a member of the CSRC (CSRC (M) (75)/12.13) thought that RLGs needed guidance on what a strategy document should contain, namely, the main policy issues, a statement of current research and its gaps, of how gaps were to be filled, the results of research commissioned and its policy implications. It would require an epidemiological framework for studying 'need'; a statement of overlaps and gaps; a statement of the economic aspects; a statement of ways in which RLGs might function - here 'flexibility was essential' - and it would need guidance on futurology and the role of scientific advisers.

At the end of a two-day meeting at York University, the CSRC underlined the dominant role of the RLGs: the RLGs were to continue to formulate and promote research objectives; monitor and if necessary advise on commissioned work; review the dissemination of results; stimulate appropriate developments, and advise on the implication for service policy (CSRC (M) (75)/12).

At the same time, however, the CSRC would make its recommendations among the relative priorities of RLGs. It made attempts, not sustained, to consider how RLGs might be encouraged to engage in futurology and apply their minds to different modes of evaluation. By May 1976 (CSRC (M) (76)/27), RLGs were invited to consider long-term requirements. Later, the CSRC determined to seek from the RLGs annual statements of their policy considerations and research requirements and oral presentations on their research strategies. It recognised, however, that to sustain such a supervisory programme it would require advice on departmental service policy and priorities.

Throughout this period, members of the CSRC noted the progress being made by the RLGs, but one member (CSRC (M) (76)/12.29) thought there should be a close examination of a few RLGs which would enable the committee to establish its own focus. The committee considered that RLGs tended to be concerned with immediate problems, whilst 'the important distinction was not perhaps so much between short and long term but between investigations which define a problem and those which aimed at identifying its causes, with a view to possible prevention.'

In wrestling with the task of supervision, the CSRC faced the problem of criteria. In 1976, an RLG review panel of four CSRC members was set up. It was to examine the plans for future research submitted by RLGs and to make recommendations to the CSRC on relative priorities.

Three strategies for the CSRC were thought to have emerged: to give priority to prevention, and establish a panel with its own funds to achieve this; to concentrate on alternative patterns of care; and to favour research leading to a more effective use of resources (CSRC (M) (76)/5.6).

The three approaches were recognised as incompatible with each other. Allocating funds to RLGs would encourage the 'Balkanisation' of research because they pursued yet other priorities which had been determined by the Department when they were set up. Some members noted that research on such subjects as prevention could take time and that priority might be given to less important issues where quicker results might be expected.

These concerns of the CSRC were arrived at intuitively. No member, let alone the CSRC collectively or research management staff, analysed areas of policy advanced by the Department or expressed through the RLGs in order to discuss the elements common to all. The emphasis on prevention and on the social construction of illness was in line with the progressive wisdom of its time. The stimulus was partly member interest but also the publication of the Consultative Document on Prevention (CSRC (76)/2, DHSS, 1976, White Paper 1977) which proposed research initiatives which the CSRC hoped 'to order into priorities' (CSRC (M) (76)/ 5.26). Attempts to engage in longer-term research planning included an exhortation to RLGs to think about it, an attempt to link with the central planning staff of the Department, and a seminar on social care held at Downing College, Cambridge in July 1977 concentrating on service provision in the field (CSRC (76)/2; DHSS, 1976; White Paper, 1977). None of these attempts was followed further in the lifetime of the CSRC.

The CSRC could find no acceptable lines for the allocation of funds. This was vividly illustrated in 1976 by the 'envelope incident' when the committee was invited to give its views on allocations. It could not do so. Discussion floundered until an Under-Secretary extracted an envelope from his pocket, on the back of which he jotted down allocations for the RLGs and other research areas, with a reserve in case the CSRC later decided upon a strategy; these allocations were agreed by the committee. The DHSS senior official had used his own knowledge and judgement in suggesting these. The scientists thought this behaviour indicated that the Department was unwilling to allow them to have a say in allocation; that 'the Department knew the allocations they intended to make all along'. The Department, for its part, could not see why scientists could not identify and impose authoritative scientific criteria on such issues.

At its last meeting in April 1978, the CSRC received the reports of two further panels. One, under Professor John Wing, was concerned with RLG plans and the other with research resources. The panel on RLG plans had consulted extensively with the RLG chairmen and some members. It identified functions which the RLG was inadequate to discharge (CSRC (78)/1): these included dealing with research units, the assessment of overall scientific standards, identifying and commissioning strategic research which might span the interests of a number of RLGs or be of a longer-term nature, and determining priorities between different RLGs and between other fields of research. A reconstituted CRSC might fulfil these functions as well as oversee the research needs of those fields lacking RLG machinery. Making research priorities required judgements on policy as well as scientific assessment. There should therefore be a newly constituted body, possibly consisting of senior departmental policy-makers and scientific advisers. Judgements on allocations were for the Department and they must be more firmly committed to this strategic customer role.

The Research Resources Panel's Report (CSRC (78)/2) was concerned with the Department's research requirements over the next five to ten years on the basis of past RLG reports to the CRSC. The panel identified a need for multidisciplinary research teams. It thought it important to involve potential contractors earlier in RLG thinking about research. It wanted to encourage units to concentrate expertise in fewer service policy fields and to exercise a tighter control over the renewal or extension of a unit's contract to ensure that the programme was directly helpful to the RLGs.

In 1978, the CSRC under its new Chief Scientist, and perhaps aware that it was coming to the end of its life, agreed with the principles underlying these reports but was uncertain about how they might be put to work. It discussed the role of the Chief Scientist and how he might be better advised. Impatience with the somewhat indeterminate range of expertise represented in the CSRC had been expressed by one member at the previous meeting. In an implicit endorsement of the unity of science he thought it best for the Chief Scientist to pick his own inner Cabinet to be appointed for a five or six year stint and to work as they thought best. The RLG's had performed good work. However, 'The CSRC was a supra-RLG body but had not been able to tackle effectively the important issues of priority among the RLGs, gaps between them and issues common to a number of policy fields. It was suggested that this was because of the absence of a supra-RLG customer.' (CSRC (77)/12.10). That was certainly one factor. We return later to the more fundamental problems of creating scientific policies.

Purposes and methods of research

In considering what methods should be used by all RLGs the CSRC questioned the purposes of commissioned research and hence its relationship with policy-making. In its discussion of methods it touched upon the importance of cost-benefit analysis, the frameworks for evaluative method, and some ways of linking the analysis of future needs to policy planning. If themes such as these could be clarified as cutting across the RLGs the committee, by insinuating them into research planning, might be able to take a strategic scientific view of client-based work.

The most zealously advanced notion was that of cost-benefit analysis (CBA) and the committee accepted that it should be introduced into all projects. There should also be 'more research into possible resource saving', whatever such research might mean. CBA was ably promoted by an economist member; resistance from other social scientists, doubtful of its overwhelming importance, was muted.

The committee made a brief and under-prepared attempt to discuss evaluation on a preliminary paper prepared by research management staff (CSRC (76)/19). It concluded (CSRC (M) (76)/12, 50, 51) that simple methods of evaluation and the requirements of 'real' customers should not be ignored. Natural and social science had different concepts of evaluation and it was unlikely, therefore, that so disparate a body as the CSRC could agree on a common approach. The committee felt that no single approach should be endorsed because the topics to be evaluated were multidimensional. Yet research management had to process requests for funds to evaluate novel organisations and services and felt the need for more clarity on evaluative criteria. Policy-makers, for example, identified the need for rapid evaluations of practice trials in social service departments which could be quickly disseminated, since RLGs did, in fact, meet such demands. Somewhat plaintively, the Department remarked that its policy staff 'had to decide, following experimentation, whether to recommend novel developments and, while research might not offer a comprehensive solution, science could surely be of some help.'

All the same, the CSRC recognised that new methodologies were needed. Evaluation should be concerned with states before and after an event. Both the research commissioned and the services being researched should be evaluated. Here the committee was stumbling across the problems faced by areas of study whose paradigms were yet weakly formulated: pre-paradigmal studies enjoy no consensus upon which evaluation can be based. In the event they concluded that each RLG would have to determine its own concept of need which was best defined in terms of the area being analysed (CSRC (M) (75)/12.15-17).

One group concerned itself with futurology. It proposed that researchers should be encouraged to look ahead to future developments. Research planners needed systematic information about social, economic and technological trends.

RLGs were in any case invited to think about research relevant to longer-term policy objectives. Research could help determine the content of policy and not be concerned simply with the evaluation of the different means of implementing policies already decided (CSRC (M) (75)/12.24 - 5).

Yet both sides found it difficult to unravel policy issues through disciplined inquiry. The Department, in a paper to the CSRC (CSRC (75)/16), analysed the problems of using research. Production took too long for policy-makers and 'the Department has tended to make the research more long term and methodological than might have otherwise been the case, with a classic statement of hypotheses and well defined and controlled investigation.' This meant that research would not help policy-makers answer immediate service problems. 'It means instead that they have to project themselves ahead four years or so and try to anticipate the issues on which research findings are likely to be valuable.' The Department needed rapid snapshots; needed reviews on the state of the art; and would help RLGs to advertise their needs if scientists could produce them.

The problems of scientists attempting to influence policy were put frankly to the CSRC by a senior official concerned with social security policy (CSRC (M) (76)/12.5, 6, 10). The scale of the social security operation was a strong influence against change. Moreover,

> much of the research proposed had been of a political rather than an objective nature. Then there was the tendency for research to try to cover the whole system rather than parts of it and the danger that research might create expectations of higher benefits, or new benefit cover, which could not in practice be met. All these factors had conspired to limit likely areas for research.

A social scientist admitted that faults were not all on one side. Social scientists had not produced an adequate scientific framework within which both fundamental and applied questions could be asked 'with due regard for political realities'. 'This pointed to a need to develop a sociology of social security.'

Another difficulty was confidentiality. Because personal and confidential data could not be disclosed outside the Department without the individual's express consent, the Department's own staff rather than external researchers should undertake research necessitating access to them. When a member expressed concern that the DHSS should determine which information could be publicly available - for much of it could allay public disquiet and fears about aspects of the system - the point was noted with a warning that matters such as confidentiality, cost of publishing and staff side objections to additional work should not be minimised.

Supra-RLG tasks

Although the CSRC declared that it would not only watch the RLGs on scientific standards, 'but also their preferences or biases about particular research styles', in practice it made no attempt to supervise RLG judgements on scientific standards. They could reach decisions on recommendations of their own advisers (CSRC (M) (75)/7.9). A second opinion would be given if asked for, but that never happened. RLGs were encouraged to put policy needs first: they were responsible 'for building up a research reservoir in the field, and in barren areas this might mean being prepared to lower scientific standards in order to foster the interests of potential researchers.'

The CSRC found it difficult to create scientific policy. Scientists can identify discrete policy problems which might yield to disciplined inquiry, and that happened in the RLGs, but members were not invited to develop science for science's sake. The Wing panel, referred to earlier, carefully distinguished between policy and scientific judgements. Policy was for the Department but it was thought that some scientific judgements could alter the present distribution of funds. They might include: What projects came up to a given standard? In which policy areas was research most feasible? Which units should be retained on quality criteria? Were there areas where units should be established to ensure a good state of research? How could topics of strategic interest to the Department most usefully be investigated? Such judgements could not be based simply on reports from RLGs, the panel thought.

Some of these judgements, however, would not be those of scientific policy so much as checking or improving on the scientific judgements of other scientists. And some were as much issues of social policy as of scientific policy. However, whether or not we follow the Wing panel analysis, there were macro-scientific issues to be tackled (CSRC (78)/1).

The CSRC did, in fact, venture initiatives in such issues as the encouragement of multidisciplinary research, in evaluative studies and in selecting the major issues of policy upon which there should be research emphasis.

In our view, however, three main groups of concerns can be considered under the heading of macro-scientific policy. There are the developments of knowledge which cross disciplinary boundaries and which call for multidisciplinary organisation. Scientists expert in single fields might, between them, explicate areas of knowledge to which more than one discipline might contribute where, indeed, new disciplines should be developed, and then specify the conditions under which such progress might be made. The CSRC scientists were generally committed to multidisciplinary advance. Their thinking, however, started with social problems which created the demand for multidisciplinarity rather than with academic disciplinary problems which should be merged into multi-disciplinary work. Given more time, they might have approached the task from both starting-points.

A similar kind of concern, capable of leading to new patterns of academic enterprise, was the encouragement of evaluative research. Many RLGs encouraged researchers to enter evaluative studies and this was an important area for advance in social policy studies. In this zone, too, however, the CSRC had insufficient resources to make much progress. The same can be said for the attempts to introduce cost-benefit analysis across the RLG spectrum and to develop some futurological capacity.

Such academic development would need to be handled by a body such as the CSRC so as not to damage the discretion of the scientists working on particular projects or within individual RLGs. The evidence is that they adopted the appropriate approach. They were concerned, for example, to promote eclecticism and liberal approaches in evaluation rather than simple and single evaluative frameworks.

A different order of macro-scientific task is that of rule setting. A body such as the CSRC could be expected to monitor the way in which the standards of science are upheld, and, where necessary, reinterpreted. The ways in which the Chief Scientist and his advisers inspected units, for example, the procedures for the evaluation of completed research projects, and the question whether academic standards might be lowered in order to get work started in unpopular areas, all such issues call upon academics to achieve consensus about what constitutes good scientific behaviour. Thirdly, the CSRC could have engaged in working out policies on scientific manpower, although this would certainly have moved it into the realm of more general policy. It touched on many of these issues, but the most important issues and priorities identified by the CSRC were those of social policy rather than generalisations derived from scientific knowledge.

The intermediate levels

The Health Services Research Board (HSRB), the Personal Social Services Research Group (PSSRG), and the Chief Scientist's Research Committee (CSRC) were all intended to take an overall view of the work of groups below them, to make recommendations to the Chief Scientist and to state priorities among the different fields of research. The Panel on Medical Research (PMR) was to establish priorities for funds transferred to the DHSS from the Medical Research Council (MRC).

The HSRB and the PSSRG were to assess the quality of larger or more difficult items of proposed research work when asked by the Chief Scientist and systematically review work in progress, particularly that carried out by the DHSS-funded units. They were to advise on the scope and quality of the resources available to the Department, whether they were to be increased, and on how to achieve a balanced use of them. In common with the CSRC itself, they were to advise on policies and priorities emerging from the RLGs (PC (73)/17.29).

Although intended to be similar in status to an MRC policy board, the HSRB had no budget or executive functions and was advisory to the CSRC rather than a board with power in its own right. Early on, DHSS officers began to feel the HSRB to be superfluous, coming between the RLGs and the CSRC, re-doing work already done elsewhere. Members began to feel frustrated: they had expected to have real influence over grants made by the Department. But the HSRB was officially disbanded upon a recommendation of the CSRC to the Chief Scientist in December 1975.

The PSSRG's activities and fate were similar to those of the HSRB. It considered reports of visits to DHSS-sponsored units and approved the research programmes of such units; it considered general approaches to research on social service topics; and considered funding for projects focused solely on social services.

The brief lives of these two intermediate bodies exemplify general points about middle tier committees. They had insufficient work to do because they lacked immediate contact with customer requirements, and could not therefore elucidate priorities. They derived their remit from the top of the system which, too, was struggling to clarify its role, and not from problems generated at the base.

Relations with the Department

The relationship between the committees and the Department takes us into the conflict between analysis of issues on scientific dimensions and the policy channels established in the Department following the McKinsey reorganisation.

The DHSS had followed the McKinsey Report in establishing a Service Development Group consisting of divisions based on client groups. This entailed assumptions about the ways in which wants, needs and services were to be analysed and policies created to deal with them.

But the Chief Scientist's resources for analysing the policy needs for research were weak and not deployed in the same ways as the policy divisions. The RLGs did not cover the whole range of departmental concerns but reflected the Department's best guesses at important but un-researched zones of action. There were serious gaps, particularly in health and primary care. As we have seen, many recommendations for priority were thematic: the prevention of illness; alternative patterns of health and social care; optimising the use of resources. These corresponded to the needs of many client groups at once and thus cut across the lines of policy and client divisions. Members were concerned with the impact of external factors on service provision (for example, community care and the role of women), with longer-term research on such issues as the changing structure of the family, unemployment, behavioural change, or attitudes of people towards services. These were background to HPSS policies rather than problems on the active policy agenda.

It was not the business of scientists to follow lines set down for organisational purposes by policy-makers. But a counter-analysis is no good unless those making the analysis being countered accept that the two must be brought together and reconciled. No machinery existed for that to happen. Despite the numerous occasions when such topics were raised, members never moved the Department into taking these questions on board, or found structure within which such questions could be handled.

There were other difficulties, too: some topics called for the investigation of roles of professional workers at a time when services were under pressure and when different groups were demanding a change of status. The Department had enough battles on its hands not to want to take on more.

Moreover, the DHSS was the largest single funder of HPSS research in the UK. Not only did the intellectual concerns of scientists lead them to expand ideas on what research the Department should be involved with, but this was done with the knowledge that there were few other sources of funds. The DHSS was invited to tackle problems no one else would. And the Department was not reluctant at the beginning to encourage creative development of its field.

The customers for advice at the macro level made brief appearances towards the end of the committee's life. Deputy Secretaries attended CSRC meetings when it was finally accepted that policy for research had a relationship to service policy. But between 1973 and 1978 the Department was preoccupied with issues so pressing that little time for non-urgent matters was left at the high levels. Its own apparatus for policy-making above the client group level was only just installed. There was a dry-run of the planning system in 1976 with a more detailed trial in the following year. It would take some time before regional ten-year strategic plans would be sufficiently reliable to highlight problems for research. 1976 was also the first time that the Department published an overall strategy for HPSS (DHSS Consultative Document, 1976). The Department had a management board but not a policy-making or strategic group above the Deputy Secretary planning meetings with the service development divisions. So time and structures available for work on overall service policy were lacking. There was no customer for the CSRC to listen to, and while the CSRC welcomed the attendance of senior policy-makers, they themselves were not in a position to contribute usefully to this kind of debate.

Disbandment and some conclusions

The CSRC was disbanded in 1978 and with it went the corporate expression of scientists' views above the level of the RLG. This was part of the reversion to the *status quo ante* in which the transferred funds were returned to the MRC, the DHSS research budget was stabilised, and the positive encouragement of customer interest in research commissioning was banked down.

The last meeting of the CSRC seemed to display a rather hopeless sense of *déjà vu*. Members sensed that the new Chief Scientist was not sympathetic to the existing structure or to the standards of work which he suspected it had encouraged. By now they wanted more action higher in the DHSS structure. The CSRC lacked confidence in its ability either to be actively useful with the RLGs or to relate closely with the policy system. In the end, it all seemed too difficult.

But the difficulties were intrinsic to the whole enterprise of securing co-operation between government and science. In establishing such institutions within the Chief Scientist's Organisation, the DHSS had two overt objectives. One was to enhance the customer-contractor relationship, by bringing together policy-makers and researchers. The second was to find means of handling the transferred funds from the MRC, which overnight considerably increased the size of the DHSS research budget. At the same time the structure was concerned with regulating scientific judgements and so was also similar to that of a research council, with grant-making committees as a first level, a strategic board dealing with a limited field as an intermediate level, and a council overseeing the entirety of research policy.

These were difficulties of function. Difficulties of relationship were equally evident. How could scientists negotiate for a place in the policy system? Co-option did not work; nor did exchange, because that which the scientists could offer was not enough for the policy-makers to be patient about the need to work out the necessary mechanisms of negotiation.

Chapter 6: The Chief Scientists's Organisation and the Research Councils: the Case of the Panel on Medical Research and Relationships with the SSRC

SO FAR, we have been concerned with the way in which government attempted to commission science through its own Chief Scientist's Organisation. But an important aspect of the Rothschild formula (Rothschild, 1972) was that customer departments would, where it was appropriate, also commission research through the research councils acting as contractors. In order that this principle could be applied to biomedical research, about a quarter of the Department of Education and Science (DES) allocation to the MRC was transferred to the health departments. This proportion was later reduced to about a fifth. No similar arrangement was made for the SSRC funds because that council was thought still to be in its infancy and, indeed, in 1982 Lord Rothschild declared that 'when one examines the work of the SSRC, there is very little to which the customer-contractor principle can be applied' (Rothschild, 1982) because even if the end-product can be defined, the end-user often cannot.

The MRC and the Panel on Medical Research

The Panel on Medical Research (PMR) was established to advise the departments on the commissioning of research through the Medical Research Council. Its brief history provided the most severe test of government's ability to influence well-established science.

The DHSS stated the role of the PMR as follows:

> The Panel's first concern and its principal task throughout are the determination of Health Department priorities in the biomedical field. On the basis of evidence presented to it, the Panel will advise the Departments* and through them inform the Council [the MRC] of the relative priorities of topics, giving high priority to topics which on one or a number of counts justify further biomedical research effort and low priority to topics where new research efforts are unlikely to be fruitful or where existing research expenditure might be curtailed. Over a number of years the Panel will develop an overall view of relative priorities, subject to periodic reappraisal and adjustment with changing circumstances, which will form the basis of the Departments' contribution to joint policy development and of the evolution of the commissioned research programme. The Panel's role in this is essentially strategic and although individual members will be involved in studies of particular topics the Panel as a whole will be concerned less with the detail of the Departments'

* The panel was to include in its terms of reference the needs of the Scottish Home and Health Department, (SHHD) as well as those of the DHSS.

> requirements in a particular area than with weighing the claims of that area against the claims of other areas for commissioned research funds. (PMR (75)/5)

This statement maximised the PMR's roles. Such phrases as 'the determination of health departments' priorities', 'strategic role', 'the Panel will develop an overall view rather than relative priorities' all implied a major input to research policy development.

There had been an agreement between the DHSS, the SHHD and the MRC which provided for the identification of notional departmental commissions within the MRC work:

> formal documents recording a commission will be in broad terms and will specify the objectives of work to be done, the fixed period of years during which the health departments undertake to pay for the work and level of payments; they will not specify the scientists or the scientific design and techniques to be used . . . the MRC will undertake the detailed definition of the individual projects, sets of which will be defined to meet the health departments' objectives; this definition will be undertaken in close but informal consultation with Health Department staff and advisers. (PMR (74)/18, Annexe 4)

The MRC was to provide detailed working papers showing the nature of the work being undertaken, and the departments were to assess the effectiveness of the system retrospectively by considering the extent to which their objectives had been met.

The PMR at work

At the beginning of the whole exercise it seemed that about thirty broad commissions would need to be vetted each year if the total of 141 notional health department commissions for the current MRC programmes were to be assessed. The commissions, on a fall-in period spreading over five years, would first be reviewed for scientific validity, significance and implications for the departments by MRC boards. The departments in turn would state their needs in those areas, and the PMR would assess the degree to which MRC commissions met them, advising the departments whether to commission and at what levels of financial support.

A commission would be selected from the part of the programme that corresponded with each MRC board and two members of the panel would review the work and appraise the commission. In the first year of the PMR's life the subjects selected for this scrutiny were to be neuroses excluding depression, the leukaemias, and sexually transmitted infections.

Members of the PMR got to work as did the departments. When the first results came to the PMR for discussion (PMR (M) (74)/3) the departments were defensive about their own contribution. Officers referred to the shortness of time available for carrying out the in-depth review of the commission on neuroses. The inquiry into departments' requirements had been less deep than was desirable. There was no established policy upon which to base a forward look and so the review was essentially a look at work in progress. The commissions were artificial constructs and considering them in isolation from related work in the MRC's and other research programmes 'compounded this artificiality'. Members, too, commented on the difficulty of conducting a meaningful review when information on relevant work in other research programmes was absent. The PMR expert who had reviewed the neuroses commission said he thought that the commission was well constructed and the title appropriate, but he found it difficult to say much about the relevance of the work in the absence of knowledge of the health departments' requirements.

In considering broad commissions on leukaemia and a special report on nutrition, the panel found it difficult to decide the relative priority to be given to a particular area of research in comparison to others. The report had been framed from a scientific point of view and was not concerned with the practical implications for the departments. Throughout, members felt the need for departmental policy divisions to state their problems more clearly. Because of the difficulty of establishing policy criteria, groups began to develop (FN (75)/1) within the panel. There were some who believed that it was important to promote scientific excellence and achievement and this might mean going ahead on all fronts at once; others believed that there had to be vigorous selection among priorities.

After a year of working it was recognised that the PMR could not systematically tackle all of the broad commissions. Instead the DHSS proposed (PMR (75/5) that the panel would select perhaps five or six subjects each year on which requirement studies would be carried out. The panel's selection would take into account not only their own views of what were the areas of priority interest to the departments, but those, too, of the Chief Scientists in the two government departments. They would also take account of the priorities expressed in the MRC stocktaking review and subjects included in the MRC programme of policy reviews.

But, as the DHSS itself later stated (DHSS Press Notice, 1980), the departments were unable to put forward ideas for new research and found themselves largely reacting to proposals from the MRC. Moreover, the DHSS's review processes unnecessarily duplicated the council's own stocktaking arrangements. From our observations, it was clear that this ambitious attempt to match science and policy evaluation was not working. Members were unhappy from the beginning. On almost every item of some of the agenda in this panel's brief life, the discussion turned to the unsatisfactory functioning of the panel. At the second meeting, the departments felt it necessary to submit a paper which referred to 'misconceptions concerning the relative importance of the different aspects of the Panel's functions' (PMR (74)/11).

The PMR was therefore abolished and, in 1977, the departments agreed a simplified system in which each year they would set out policy and service problems to which biomedical research might be applied (CSRC (77)/9). The departments would prepare an annual statement of policies and priorities which 'will act as a trigger mechanism in the formulation of the commissioned biomedical research programme.' The CSRC would be involved in the formulation of the statement. This would accord with the DHSS's analysis of the 'burden of illness'. The MRC would then suggest how its research programmes might relate or be developed in response to NHS needs, and in this way it was hoped a relationship which recognised the distinctive role of the two parties could be brought about. These revised arrangements came into operation in April 1978. Not long afterwards (in 1980) the transferred funds were handed back altogether.

Why did this attempt at collaboration between the health departments and the MRC come to nothing?

MRC view of how science happens

The Secretary to the MRC, Dr James Gowans, made explicit in evidence before parliamentary committees his assumptions about how good research is created. He went out of his way to explain what he considered to be the peculiar difficulties of biomedical research. The MRC had a very wide remit. It undertook, Gowans said, 'research all the way from basic mechanisms to very applied and practical work from for example, molecular biology on the one hand to toxicology and nutrition research, and industrial injuries and burns and so on.' (PAC, 1979, para 1288). At the time of his evidence over fifty clinical trials were currently in progress. Most important

> it has under its surveillance all the work, right from the basic to the applied end; in cancer, for example, we would consider it essential that the Medical Research Council

> should start with basic molecular biology...and go right the way through to trials of new chemicals and radiation therapies. That spread implies...that if anybody has a large sum of money to spend with the MRC, they really must look at the whole of the MRC's programme in order to spend it.... There is very little sense in abstracting out of it a little fraction of it and saying 'we will concentrate on that.' [it follows] that the concept of the project ... is scientifically a rather difficult one to isolate in cost terms.

Doctors and scientists want to work on the killing diseases.

> There is a public mood for working in these areas. But they are all areas where in principle one does not know the answer.... One is waiting for the next good idea. Since one is waiting for the next good idea, it is work on a very long timescale, and it is not work that lends itself to targetted research. You have to invest in good ideas and good people and it is impracticable therefore .. . [here he quoted Rothschild] 'to define the objectives closely enough for the placing of specific commissions'. . . you are waiting...for the next bright man, and the prime thing is to invest in that. (ibid)

The key discoveries - and he referred to three Nobel prizes given for this kind of work - which have had an enormous impact on human welfare have all been made by accident. These were, he thought, the points that made for difficulties in the Rothschild partnership.

How did the MRC believe it reacted to social needs? The MRC responded to all the pressures under which it found itself. The key pressure was that of the biomedical community.

> That is the community of doctors and scientists who spend their lives either looking after sick people or working. . . . We hold the mirror up to the biomedical community. If they are interested and concerned about a particular theme, we tend to reflect it. That is the passive approach. More actively, the Medical Research Council feels that it should be entrepreneurial where it can and should try and fill in gaps, if you like, where there is social pressure but no activity in the scientific and biomedical community. That is the difficult one. The difficult one is matching the bright people to the difficult problems, because they are not necessarily working in areas where there is considerable public concern - suffering, morbidity and so on. (ibid, para 1292)

Here Gowans was implicitly endorsing forms of steerage, but by research councils rather than by government.

In response to a question, Gowans agreed that the MRC might accede to the most persistent lobby. But, it was implied, the lobby was a knowledgeable one. The MRC had 150 or so doctors and scientists serving on boards and council at any one time. There was a flow of grant applications, which led to a consensus view of where the feasibility was. Yet there were areas of less activity than one would like in view of the extent of morbidity and suffering in the community. It was difficult to get medical research in, for example, common eye diseases, obstetrics and gynaecology, in dermatology and dentistry. These were not areas that interested the scientific community for reasons bound up with the nature of medical practice and career systems. The MRC undertook training programmes in these specialities. It might send individuals abroad or build an institute or a little unit 'around a good man'. These were MRC initiatives not based in universities.

Again, in response to challenge, Gowans thought the health departments had every opportunity to state customer needs.

Gowans' evidence was supported by Sir Patrick Nairne (ibid, 1289), the Permanent Secretary to the DHSS, who presided over the later return of transferred funds to the MRC. He thought that the Rothschild policy had extended and improved the interaction between the Department and the MRC. There were personal contacts between the Chief Scientist and the Secretary to the MRC, and liaison had been excellent. But the broad commissioning approach turned out to be burdensome and 'strictly unnecessary'. Instead, the DHSS now analysed the burden of disease which was the basis for a 'kind of annual running partnership'. A few small specific commissions had been made on the MRC. These were on ultra-clean air, on influenza vaccines and on drugs for whooping cough.

Arguments such as those of Gowans eventually convinced the health departments to release their hold over MRC funds. They demonstrate how leading scientists differed from the DHSS on ways of getting work done. They would note what the policy-makers analysed to be the burden of illness. But, essentially, they would act responsively to the perceptions of their own kind, scientists and practitioners, rather than to those of policy-makers accountable to the wider public.

The MRC could not only make a good case in public but was also convincingly efficient in its internal working. It was already reviewing scientific policy within a cycle of about five years. Each year it took stock of the overall balance of its efforts and its priorities in the allocation of research resources. Its priorities included health problems which, because of their social importance, the council wished to keep under continuing review: 'by means of ... regular scrutiny the council aims to seize any opportunity that occurs to support research likely to contribute to the solution of these problems.' (MRC (74)/ 323).

The departments' problem

By contrast, the departments knew they needed more help, but could not confidently state their needs. And to us, as observers of the process, their comments on the broad commissions and their analysis of other research not falling within the broad commissions, represented a strenuous effort which might have led to serious policy analysis had it been given a chance to develop. The departments did make statements of priority with suggestions on what should be reduced, what should be maintained and what should be increased.

In one of their earliest papers the departments stated the essential task of the panel as 'perhaps to identify the questions in medical practice that the MRC from the nature of its approach would tend to miss', questions, for example, relating to the undeveloped specialities, general practice, questions of alleviation as opposed to cure, scientifically un-exciting questions amenable to an empirical, tactical approach (PMR (74)/3). Quite soon after the Rothschild Report proposals were accepted, a joint MRC-health departments working group produced arrangements for co-operation in the field of biomedical research. Its statement of the factors determining the arrangements for research commissioning bore marked resemblances to Gowans' evidence to the PAC six years later. Because biological processes with which the research is concerned are highly complex, it was rarely possible to define the course of a research programme in advance. Ignorance of the physiology and pathology of the human system is still profound and progress must depend upon broadly based programmes of research. This kind of 'strategic' research might be aimed at intermediate scientific objectives and contribute some evidence and knowledge to the solution of practical problems. It did not, however, attempt to reach those objectives by means of a single programme of research planned in some detail from the start. 'It is this type of research in particular which is unpredictable and difficult to cost while, at the same time, the health departments have a substantial need for it.'

> The arrangements must therefore give them an effective voice in the selection of priority subjects for strategic research and enable them to ensure that openings giving reasonable promise of early practical benefit are not overlooked and that the planning of service and research are coordinated. The Departments will of course require an appropriate balance of 'strategic' with 'tactical' research. The two categories will tend to demand different arrangements for operation between 'customer' and 'contractor' ...The primary motive in many projects and programmes is ... to increase knowledge rather than to meet short-term practical needs. Fortunately a good deal of work in biomedical research is of great interest and importance on both practical and theoretical grounds and for both short and long term reasons. (Joint MRC/Health Departments Working Group paper, 1973)

Later on (October 1977), when in retreat from PMR, the health departments stated more trenchantly their requirements and priorities in the field of biomedical research. They were concerned, they said, that the treatments and other procedures adopted in the NHS should be the most effective and efficient available in the circumstances of practice. This was particularly important when major investment decisions on plant and equipment were at stake.

> There are...many diseases for which satisfactory prevention or treatment is not available. One response to this is to give priority to the search for the unknown mechanisms of the disease and this is perhaps the appropriate response by medical scientists using the Council's own funds. The cost of the diagnosis and treatment of these diseases is great and there is reason to think that relatively ineffective treatments are often given to the disadvantage of patients and of the public purse alike. Similarly, the place of some diagnostic procedures is not properly established. Continuous advances in palliative and supportive treatment face this and other countries with the threat of being unable to afford all, even clearly effective, available treatments. So far as the commission funds are concerned, therefore, the health departments give priority to the exploitation of existing knowledge by the development of improved methods in prevention, diagnosis and treatment. Particular attention to any possibility of less expensive, equally effective alternatives to existing methods is obviously desirable. Particular priority is also given to research into the relative effectiveness of existing alternative methods and regimens, as, for instance, by controlled trials of treatment. Problems faced by all types of health care staff, whether or not they work in research oriented teaching hospitals, deserve equal attention, in selecting topics for research.

The document continued, after this gentle smack on the head for elitist scientific medicine, to enunciate the doctrine which had also been advanced to the CSRC (see Chapter 5): 'the health departments hold that, other things being equal, priority should be given to common conditions and that research expenditure on different diseases should be in rough proportion to their relative burden on the community.' (CSRC (77)/18). Here they referred to the index of the relative burden of different disease groups derived from the paper by Black and Pole (1975). Whilst acknowledging reservations about this index, 'nevertheless the health departments hold that broad indications of priority can be drawn amongst the relevant parts of the commission programmes.' They therefore asked the MRC to comment on how the commissioned funds were spent by reference to the index and to provide conveniently available information on related research in the country.

The health departments at different times acknowledged conflicting positions on medical and health research. First and last, they accepted the largely internalist view put forward by the MRC in their earlier joint document (Joint Paper, 1973, op. cit), and later repeated by its Secretary before the PAC, that much of the best medical science develops from the knowledge and interest of individual scientists and practitioners which are presented spontaneously to the MRC. The MRC could clarify a general professional consensus. Secondly, they maintained that within that work, which would be driven by intellectual curiosity or by knowledge derived from working with patients, there were still areas of social and individual need that were not being tackled and these the departments would identify with the help of MRC-style scientists working on the PMR and through their own strengthened links with the MRC. Thirdly, the DHSS would continue to have its own health services programme to include studies of health care systems, patient management, clinical practice and social and community medicine.

An impossible enterprise?

It is difficult to be clear whether the problems of collaboration between the departments and biomedical scientists were intrinsic to the task or whether they arose from preconceptions of role and status. To some extent the task was an artificial one. The scientist members of the PMR were required to step beyond their own expertise in order to help the departments identify areas where work was not yet being performed or where priorities could be reasonably changed. The departments hoped that the scientists would extend MRC criteria to fields and questions at present neglected. They were thus depending upon them for social and scientific judgements. But, in the event, they were diffident about forcing scientists to face the issue of how medical research might connect with problems of practice and service. So they limited themselves to asking questions on the relevance and level of support to be given to the broad commissions. There was thus a built-in assumption that biomedical research was in some way separate from health service problems and that, in consequence, neither the CSO nor policy divisions in the DHSS had a contribution of their own to make.

Members of the departments sometimes found themselves stepping beyond their expertise of policy and practice to apply scientific criteria. For example, medical officers developed arguments on how research on neurosis ought to be carried out, including proposals for 'tactical research' which would entail trials of treatment methods in operational settings as part of the more comprehensive or strategic programme of research. This sort of judgement could not be made by scientifically innocent contributors to the discussion.

A second set of issues centred on decision-making. There was a perceptible difference in atmosphere between the PMR and such bodies as the Small Grants

Committee (see Chapter 7). The Small Grants Committee took decisions and looked and sounded as if it did. In effect, the PMR also took decisions, but that was not how it felt, and members were uncertain of their impact. It made recommendations about research commissions which could not be rejected or ignored: it was virtually inconceivable that the health departments could commission a project thought to be scientifically doubtful by the panel. The MRC membership thus had both a negotiating role, for its members inevitably voiced the claims of biomedical research to government policy-makers, and also exercised a *de facto* veto.

But the machinery was cumbersome - at one meeting which we observed over twenty people attended for the whole day, a majority of whom were officials of either the MRC or the two health departments - and members became impatient with it. This impatience was exacerbated by the uncertainties of the departments in their relationship with the scientists. Officials tended to bring issues to them in case they were offended at not being consulted. Their reward was irritation at being consulted when effectively the decision had been made and was anyhow thought to be entirely within the departments' own competence.

Thus, although the panel could have developed power because it could make recommendations leading to MRC commissions, it did not. Its difficulties were enhanced by the heterogeneity of its scientific membership. It represented a diversity of disciplines, only some of which were directly related to each other. It was hard to generate a common purpose. The departments wanted to be good customers. Their ministers, administrators and professionals had long been anxious that good science should be applied to the needs of patients in the NHS. It did not attempt to sabotage the MRC's ways of working for 75 per cent, later 80 per cent, of the Council's fund still remained, for virtually freely determined distribution to scientists by fellow scientists. But functioning became impossible and the PMR was quickly put to death.

The story of the PMR exemplifies the impermeability and authority of a Medical Research Council grounded in an internalist view of science, and in the notion of indivisibility between basic and applied research. But its impregnable position rested too on the almost paradoxical claim to responsiveness to social norms and pressures through the work of the medical profession. It was willing itself to steer scientists, where social needs conflicted with medical science's own hierarchies of reward.

The departments could not assert the authority to breach this position. The reasons were social, epistemological and institutional. The Department members were deferential to the scientists on the PMR who represented the peer group with the most deep-rooted influences upon the professionals in the Department. The policy-makers felt that they had no distinctive contribution. The policy criteria which they eventually produced were too broad for scientists to use. The 'commissions' within which the MRC's work was framed distorted the complexity of the research, without enabling either the scientists or the policy-makers to cross the divide between scientific and policy priorities.

Finally, the rewards for success remained unclear for the scientists. The mechanism of the transferred funds represented a threat to a powerful scientific institution. If the DHSS could not substantiate the connections between biomedical science and health services problems and practice, what was to be gained by collaboration? Science might be weakened without any compensatory advantage to policy.

DHSS and SSRC

The relationships with the SSRC were of a different order. The SSRC was one of the two research councils not required to transfer parts of their budget to government departments. The SSRC was in 'its infancy' (Rothschild, 1971) - it received its Charter in 1966 - and it was implied that a change at this time would be helpful to neither the SSRC nor government departments. But the SSRC could see the writing on the wall and, on its own initiative, attempted to ensure that its research programmes incorporated strong elements with policy relevance.

At about the same time as the establishment of the Chief Scientist's Organisation, the SSRC set up a Research Initiatives Board which created panels for research issues of policy relevance, such as health studies, smoking and some other issues that could be thought to come within the zones of health and social welfare. The DHSS commissioned the council to produce a programme (£0.75 million) of work on transmitted deprivation which produced impressive results (Brown and Madge, 1982).

Whilst the SSRC was certainly under pressure from both the Advisory Board for Research Councils and its own chairman to take initiatives in policy-related research, social science research contributions to policy and practice were never considered as obviously beneficial as MRC research into causes of clearly defined disabilities. The

1979 White Paper reviewing the Rothschild Report and its implementation (*A Framework for Government and Research*, 1979) in fact noted that few government departments felt the need to use the SSRC as an intermediary with researchers as they would use the other research councils.

The SSRC Panel on Health Studies awarded contracts of £250 000. Whilst it was intended to advance studies based upon independent social science research with particular emphasis on multidisciplinary work, the panel had DHSS members and it was hoped that work relevant to the NHS would result. The issue for the SSRC was whether it could hold to its independent position whilst being useful. Its panel started by verifying the range of concepts of health that emerged from both medical and social science literature (Stacey, 1977). It noted how the biomedical or clinical model was still predominant and was sure it would remain important. But it was thought that health systems should increasingly embrace concepts derived from social work and community care as well as those centring on specialist clinical help given in hospitals, based on biomedical research. Different concepts of health produced different systems of health care, with different bases for evaluation. They in turn led to different assumptions and knowledge about outcomes and produced different decisions and policy-making structures and different divisions of labour. The panel thus strove to build a catwalk between conceptualisations and practical issues for policy-makers.

Somewhat later (1981), the Secretary of State for Education and Science, Sir Keith Joseph, invited Lord Rothschild to scrutinise the work of the SSRC. The charge to Rothschild contained an assumption that had just been eradicated in government's dealings with the MRC, that if some of the SSRC's work was useful, those to whom it was useful might meet the bill.

Lord Rothschild's Report (Rothschild, 1982) lent no encouragement to any attempt to demolish (which Rothschild thought would be 'vandalism') or reduce the SSRC or its work. The contrast, however, between ministerial deference to the MRC, with its essentially imperial and internalist view of the role and rights of the scientist, and the harassment of the SSRC as it tried to mount policy initiatives must be noted.

A further contrast can be drawn between the MRC and the SSRC in terms of their response to the original Rothschild challenge to be relevant. The Gowans' version of biomedical research was that of progressive, accretive and self-reinforcing systems of thought in which able people developed formulations that could be tested and added to the stock of knowledge. The SSRC initiators, by contrast, assumed that health and health care were capable of multiple definition from different stand-points. Definitions might conflict and present a range of normative preferences. By the early 1980s the Government eventually supported the internalist and more aloof perception rather than the more eclectic views put forward by the SSRC although they were clearly more capable of steerage by policy-makers.

An evolving relationship

The health departments tried to work out new arrangements with the research councils. This was accompanied by change within the councils themselves. In the SSRC, the power, institutional, intellectual and methodological, rested with the subject committees until radical modifications were made in 1981. Multidisciplinarity was intended to emerge from the new scheme. But the problem of receiving and responding to government customers remained. SSRC initiatives to meet policy needs were made by concerned social scientists. In our view, they lacked not so much the voice of the customer as his detailed working commitment and co-operation. The MRC had the easier task because it enjoyed the confidence of a single, and supremely self-confident, profession strongly represented within the departments. It was not clear that health services research within the MRC would be given enough room by its academic structure and intellectual methods. The whole system of brokerage and of enunciating policy needs, so painfully built up, was not brought to fruition, within the DHSS. The Chief Scientist of the time made it clear throughout his period at the DHSS that such matters were not his priority. His central concern was with the quality of health services research and what he saw as a prerequisite: a strong national base. He thought the Department 'unsuited to accept the total or even the major responsibility for [this] base' (ID (80)/28).

In 1980 a new concordat was drawn up (DHSS, SHHD, MRC, 1980).

> The Health Departments and the Medical Research Council agree that the commissioning funds will be transferred back to the DES Science budget on the understanding that the Medical Research Council will continue to meet the needs and priorities of the Health Departments in the Council's programme of biomedical research, and will also undertake to mount and manage, in partner-ship with the Department of Health and Social Security, some health services research on the basis of agreed administrative and financial arrangements.

The departments' formal statement of requirements would be considered by the MRC. The DHSS Chief Scientist and two Chief Medical Officers would continue to be members of the MRC. They, together with the Scottish Chief Scientist, would also be members of the three MRC boards to each of which the Department would nominate three independent scientists. The MRC would promote more health services research. It was hoped that the health departments would maintain a close interest in the biomedical research undertaken by the MRC and participate in the work of the council. There would be an annual meeting at an appropriately high level. In advance of it, the health departments would circulate notes on matters they wished to bring to the council's attention. The council would circulate a note on its new and anticipated scientific developments which could affect the health services, together with a commentary on work specifically requested by the health departments. The health departments might continue to commission biomedical research elsewhere, funded from their own research budgets. The MRC would, as opportunities arose, engage in health services research to a greater extent than at present in MRC units and by grant support to universities. The aim was that over time the base for health services research provided by the MRC would be increased 'so that the Council may also undertake commissioning of such research in a customer-contractor relationship with the DHSS.' The DHSS would put proposals to the council about the areas in which the Department would wish to see provision made by the MRC. The machinery for the customer-contractor interaction would 'in due course, be reviewed by the DHSS and the MRC'. A health services panel was later set up.

Problems of connection must have remained. It was not clear that the MRC had fully grasped the significance of health services studies. One example of the expected division of functions was recorded, without apparent irony, from an earlier discussion between the Chief Scientist and a senior officer of the MRC: 'epidemiology would be acceptable [to the MRC as a subject for research] if the objective was to test a hypothesis; but the collection of data upon which to base hypotheses would be for the DHSS' (DHSS MG (78)/9).

The encounters between the DHSS and the research councils involved three systems of different knowledge, authority and power. The MRC represented an elite, bounded scientific organisation, the coherence and authority of which were underpinned by a powerful profession.

Dominant concepts of health and organisational structures derived from them were shaped by this profession. The SSRC represented still emergent and competing conceptions of social science. Its institutionalisation within higher education was relatively recent, and its epistemological, social and political authority uncertain. It proposed multiple concepts of health, some of which challenged current basic assumptions. The Department's commissioning machine contained a combination of policy-makers and applied scientists, with both medical and social service orientations. It would have been difficult for the Department to support initiatives representing critique and alternative analyses in the face of the well-established authority of excellence in the MRC and its own tribalism. It did attempt to establish connections with the SSRC on issues more self-evidently belonging to social science. But social science research challenges much more directly than medical science research simple enlightenment or instrumental models of the relationship between research and policy. Alternative or counter-analyses or interactive models of that relationship entail issues for both researchers and government which neither the SSRC nor the Department seemed ready to take on board. If policy-makers as individuals are willing to sponsor research that challenges the *status quo,* the implications for the policy machine might be far more complicated.

Government then failed to determine the range of science that it needed, or to work at the relationships between policy and science that an extended range would require, or to analyse the mix between 'free' research financed through the UGC or the research councils and commissioned research which responded to the needs of society and the policy-machine.

In consequence, exchanges between government and science were imbalanced. What would have been required to make them work was insufficiently thought through. Exchanges thus occurred predominantly between individual scientists and particular commissioning groups within government. The need for a negotiative exchange was perceived weakly by elite scientists within the MRC who knew that they could beat the bureaucrats by appeals to the political system above their heads. The SSRC's position was different but its currency was of more uncertain value.

Chapter 7: Research Liaison Groups and the Small Grants Committee: Two Contrasting Systems

Research Liaison Groups

IF THE CSRC and its intermediate boards necessarily dealt with abstractions, the Research Liaison Groups (RLGs) directly faced problems of policy and practice. They represent perhaps the clearest commitment of the Department to the customer-contractor principle. Between October 1972 and March 1976 eleven groups were set up to provide a forum in which policy-makers, research management and external scientific advisers could work together to foster research. Research management wanted more from the policy divisions than passive agreement to research projects: it wished them to identify with scientists those issues where science might help to specify and to fill gaps in knowledge. To that purpose, the RLGs were 'given' to the policy divisions. It was they, the 'customers', who serviced and chaired them. For the first six years the groups worked within common terms of reference but were left to determine their own methods of operation and to create their own patterns of interaction between the representatives of government and science.

The history of the RLGs falls into two main periods: the first six years during which relationships between policy-makers and external advisers were built up and strategies and procedures for the commissioning and review of research established. There was then the period following the appointment of a new Chief Scientist in 1978 with an executive rather than advisory role. His review of the research management system was to impact strongly on the RLGs.

Range and terms of reference

In origin, the RLGs were an experimental and temporary device set up in policy fields that were thought to be under-researched. Once these customer requirements were established the RLGs could be replaced by others in different fields.

Lack of manpower and funds and the difficulty of dividing rationally the whole departmental field meant that RLGs were never to cover the range of the Department's research interests. The mix of rationales on which they were based can be seen in a glance at the list: mental illness, mental handicap, forensic psychiatry, children, the elderly, physical disablement, homelessness and addiction, nursing, reproduction and allied services, the local authority social services and supplies. Particular client groups or social problems stood alongside service sectors and a profession. Boundary issues were a continuing and shared concern of all groups, for the most part dealt with by *ad hoc* and discontinuous solutions.

The RLGs' terms of reference were fivefold: the formulation of research objectives; the promotion of research to meet those objectives; the development of an overall research programme to serve the RLGs' needs; monitoring research in progress and revising commissions where necessary; reviewing and acting on the results of research (IM (73)/2). Research management issued broad guidelines about how primary responsibilities for these tasks should be allocated between the component elements of the RLGs: the policy divisions, research management and scientific advisers. It assumed a collaborative mode of working. This was important, not least because of the complexity of RLG membership. There were external advisers from a range of scientific disciplines and extensive departmental representation: administrators from policy divisions and research management, professionals (doctors, nurses and social workers) with responsibilities either in a policy field or in research management, and other departmental specialists from the statistics division, the Economic Advisers' Office (EAO) or the Department's own in-house research team. The Welsh Office represented another interest group, and from 1978 there were also external service advisers from the field authorities.

Such membership created two potential problems. Meetings were likely to be large, and responsibility and expertise sometimes blurred. RLGs also had to start work without a financial framework. Existing commitments to research units and programmes varied widely between policy fields, but all RLGs were told to formulate their objectives and needs without reference to financial constraints. Later, each RLG worked within a budget to be based on a three-year plan. In the event, the economic uncertainties of the second half of the 1970s were clearly reflected in the fluctuating budgets from 1976 onwards.

Ways of Working

So how did the RLGs begin to work? The device used effectively by the children's RLG was the *ad hoc* working party where the bulk of their collective work was done. One of the apparently most coherent divisions, nursing, established three permanent sub-groups on practice, service and education but left coordination of their work to the parent RLG. Physical disablement (at first called physical handicap) adopted a similar strategy.

Different approaches to managing the field of operations were gradually reflected in patterns of meetings. The relatively small (although about twelve civil servants attended in addition to external advisers) RLG for mental illness met together on average five times a year and ran a number of working groups. The children's RLG had, by the end of 1975, settled to a pattern of only two meetings a year of the main group at which the work of the working parties was brought together. The RLG for the elderly met three times a year, with heavy input from the policy division in between.

In their early stages, RLGs attempted to formulate objectives and a strategy for promoting research within a planned programme. Many began with an encyclopaedic approach, with officers producing lists of the research which the DHSS had funded in their areas of interest and which members then attempted to relate to departmental planning statements. They tried to map the entire range of potential research topics relevant to the policy areas, in one case by circulating the academic field. These were important essays but the work required to bring them to fulfilment would have been vast. The limitations of systematic and comprehensive approaches were quickly borne in upon groups. One, which spent three meetings discussing a highly systematic paper from one of its advisers on how a policy issue might be broken down into researchable questions, was told decisively by its chairman that such an approach to the whole field was impracticable (RLG (CH) (75)/16). The exercise demonstrated not only the cost in time but also that research could complicate policy issues. The same group found that while advisers and policy-makers (for quite different reasons) agreed in principle on the most desirable priorities for research, these did not constitute a politically feasible agenda: too large a proportion of the client group would have been neglected in the short term (RLG (CH) (M) 1975).

Another group, homelessness and addictions, in reporting to the CSRC in 1978 (RLG (HA) 78/10) noted its problems in devising a research strategy because of the incoherence of its field and the uncertainty of its defining limits. However, it succeeded in attracting more researchers and commissioning more projects than RLGs in policy areas where research was more established or specialised or the field more coherent. The strategy documents on mental illness and mental handicap were comprehensive and based on a more systematic conceptual framework. Yet, in their early versions, they remained at a level of generality which did not tell researchers clearly what was wanted.

One group did make a systematic approach work. Customers from the branch for the elderly presented to the RLG an annual list of priorities, ranked high, medium and low and long, medium and short term. The RLG went through the list, selected those which it thought most important and suggested researchers who could be approached to undertake the work. The policy branch then produced a paper on each topic, stating in some detail why the information was wanted and what types of research were being looked for. This group thus accepted that priorities had to be those of policy and not of science. Records of its meetings show the pressure upon this group to contribute to policy, for example, on the politically sensitive problem of hypothermia (RLG (SHB) (M); FN (76)/1).

Finding direction and focus

The pull between the scientists' hope of systematic frameworks within which ill-defined fields of research could progress and the needs of policy-makers to make quick but authoritative responses to policy issues can be seen in the workings of some RLGs. But the publication of government commissioned reports, the activities of select committees and the pressure on central government to provide leadership in policies for particular client groups such as the elderly, the physically disabled (the first Minister for the Disabled was appointed in 1974), abused children and battered women (sic) provided focus and direction to RLGs. Policy-makers concerned with the elderly had clear broad policy objectives: the sustainment of individual independence, priority of community care and the need to maximise the efficient use of resources. However, they accepted that implementation depended upon a deeper understanding of needs and of obstacles to family care, community care and self help as well as upon the evolution of alternative patterns of care. Policy-makers began to talk of the contribution of research to policy development as well as to policy implementation.

Attitudes of policy-makers to close collaboration with members of the scientific community constituted one important variable. Some were at first wary of and even hostile to a group of people who, they thought, might regard challenging government as their main function. In some cases their attitudes were well founded. Some advisers certainly hoped they could both challenge and promote significant change. Other policy-makers, while supporting the development of research and perceiving it as an instrument of policy implementation, considered its potential contribution to policy development to be negligible. They gave scientists space in the RLGs but maintained their distance from them. Others again wanted to draw their scientific advisers in but were unsure at least initially how best to use them.[*]

But the existence of bounded fields of work, with, as time went on, their own budget, and with urgent policy problems to solve gave a powerful impetus to corporate identity and activity. The achievement of 'boundedness' derived from different sources. The nursing groups pursued a research policy laid down before the Rothschild machinery and shared a professional framework with some of their external advisers. This made it possible for them to maintain control over a wide remit. The Department, through the RLG for nursing, was almost the sole source of support for nursing research and was determined to make nursing a research-based profession. The physical disablement group had succeeded in dividing a vast field into client group or specialist interest areas with which members could identify. The children's group focused on quite specific policy issues to which its small working parties of policy-makers, advisers and medical and social scientists working together became strongly committed.

One criterion of effectiveness by which some early judgement of RLGs could be made was the amount of work they commissioned. There was some remarkable progress. In particular, between 1974 and 1978 the physical disablement RLG increased its share of the HPSS research budget from 1 to 9 per cent. By 1977 it had commissioned 27 projects at a total cost of over £700 000, almost twice as many as the next most productive group, the RLG for the elderly. In addition the physical disablement group was supporting research in the funded units at an annual cost of over £400 000 (IM (78)/7).

Each RLG established its own priorities and had its own problems. Some concentrated most of their resources upon units or programmes either to promote specialist centres in a relatively new field or under the constraint of inheritance. Others, like those for homelessness and addictions, and the elderly, whose inheritance was meagre, devoted more money to individual projects. Most worked by reacting to spontaneous proposals rather than by actively commissioning work. And the time consumed in negotiating proposals with the applicants in order to sustain

[*] Interviews with customers and research managers in the earlier years reveal attitudes that were to change greatly when attempts were made to reduce RLG activity in 1979. (Cf. reservations in ID (76)/2, 3, 6, 7, 9, 10, 11, 13, 15, 18, 19, 20, 23 and 36 with the response of RLG chairmen, incorporated in 'Proposed reduction of activity by RLGs: note of Chief Scientist's meeting with RLG chairmen and OCS lead officers on 19 August 1979', IM (79)/21).

scientific standards and to bring projects into line with RLGs' own priorities was to become a source of serious anxiety in the Department. These early efforts were, however, important to RLGs not only because of the problems of attracting good researchers to policy-relevant research but also because they could thus clarify priorities.

Finding Researchers

Anxiety about the scarcity of researchers who could be relied upon to do good policy-relevant research extended across RLGs. Some found themselves crying in the wilderness: the reproduction and allied services group, with projects as apparently topical and specific as those in the homelessness and addictions group, had found little response by 1977. Others found their resources were eaten up by the funded units where the objectives and sometimes, they felt, the quality of work were ill-matched to their own priorities and perception of the task. The story of two of the more successful RLGs to some extent revolves around experienced, far-seeing and determined leadership from the advisers in co-ordinating, focusing and gradually redirecting existing research resources. Other advisers, although committed to researcher freedom, were frustrated by their inability to engineer a closer match with research unit priorities, and by units' involvement in long-term commissions whose value had been reduced by policy changes in their fields of work.

RLGs, as the embodiment of the Department's commitment to the customer-contractor principle, inevitably had ambiguous attitudes to units. In 1972/73 the units and rolling programmes represented something like two-thirds of the total research budget of nearly £7 million and their independence and the assumptions on which they were established gave little room for manoeuvre towards new research priorities. Of the 69 units or programmes in 1972/73, only one-third related directly to the broad policy areas of RLGs; and of these, several were concerned with specialised or relatively narrow topics. This was not surprising or wrong in itself. The DHSS had deliberately planted RLGs in the least developed research territory. But the work of matching the aspirations of existing units to those of the Department, as well as cuts in the departmental budget, inhibited the build-up of alternative research resources. Units embodied the Department's priorities of the previous decade. Yet RLGs could not deny the importance of the areas covered by them or their contribution to establishing a strong base in fields of research on the edge of the remit of the research councils. A few new units were therefore created by RLGs and there was advocacy for more. The National Perinatal Epidemiology Unit was strongly supported by the children's RLG and its scope was influenced by the recommendations of an RLG working party on the first year of life.

RLGs adopted a variety of approaches to the task of seeking new researchers. The most systematic and long-term strategy was that of the nursing RLG. Its aim to establish nursing as a research-based profession was pursued through programmes of education for research, through research resources centres in funded research units, through dissemination policies and through the award of the nursing research fellowships.

Social work, by contrast, had no RLG of its own but the need to develop social work research was pursued, notably by the mental illness RLG, one of whose advisers was a widely respected social work researcher. She and the other advisers in the RLG, who were until 1977 all psychiatrists, worked hard, but in the early years they made little headway. The records of their meetings point to some possible reasons. They were determined that the range of researchers funded by the Department should not be increased at the expense of quality. Their exhortations to inexperienced researchers to apply were hedged about with provisos about the quality of supervision they would need. They tried to encourage senior and experienced psychiatrists to free themselves to lead research projects. But as time went on, and the expertise of the advisers broadened to include sociology, psychology and economics, doubts began to be expressed as to whether the quality criteria being applied, centred as they were on rigorous control, were compatible with the variables appropriate to social work research. Some completed research in the psychiatric tradition judged by the RLG to be of good quality was also considered to be of limited use. Members felt they needed a more multi-disciplinary approach (RLG (MI) (M) 1978). But they then were faced with an unresolved scientific problem: the conditions for and criteria of good multidisciplinary research. They had to wait upon 'finalisation' in the scientific community or to be content to define their needs in terms with which the scientific community could currently cope (Weingart, 1977).

Other RLGs, such as those for mental handicap and the elderly, tried to solve the problems of researcher resources by another route: encouraging researchers established in other fields to switch their interests. This, too, brought limited success. The response was small and sometimes researchers from different fields wanted to introduce values or concepts which conflicted with those of the RLGs (RLG (SHB)

(M) 1975). As RLGs recognised, they were asking researchers to do far more than simply transfer a set of techniques.

The search for research resources was a common preoccupation across RLG boundaries. Common themes also emerged in the kinds of knowledge which RLGs sought as they began to work

Range of disciplined inquiry desired by RLGs

Many RLGs were concerned about the coverage and quality of the Department's data base. Some started in ignorance of the incidence and prevalence of certain conditions, of the characteristics and needs of certain client groups, and of the extent and types of services being delivered from some organisations. Survey work formed an important part of the early strategies of a number of RLGs. Others inherited sources of data, such as cohort studies and case registers, which were not as valuable as might have been expected. Sometimes they excluded categories of information considered a priority by advisers; and some geographically-dispersed studies provided data which could not be linked. One adviser to two RLGs worked very hard to promote the refinement and co-ordination of funded case registers.

Literature surveys of existing research were considered important but sometimes RLGs had to learn the limits of their usefulness. RLGs in 'new' fields of research, such as nursing and local authority social services, wanted to establish central and reliable sources of research information, which they felt to be fundamental to the build-up and dissemination of research in the field.

But there was also pressure to use research as a tool for more effective intervention. An RLG noted in 1977 that if research is to provide useful guidelines for policy, studies must move from the descriptive and classificatory to the experimental and evaluative. The pressure was felt in both new and more established fields and not only by policy-makers. A scientific adviser initiated a shift in emphasis in the strategy document on mental illness from epidemiology to treatment research (RLG (MI) (M) 1974). But while all RLGs regarded research on intervention as a high priority, the difficulties of promoting valid work were formidable. There were linked but distinct issues: the locus of responsibility for service development and experiment within the Department and between central and local government; the relationship between research and development; methodology and criteria for assessment of evaluative research and action research into policy and practice intervention.

At an early stage some RLGs found that their scope for developing a research strategy was limited by existing commitments, many of which were, in fact, to service development rather than to research. The CSRC was asked to study this problem, not because RLGs or their scientific adviser members objected to evaluative or action research linked with development, provided (as had not always happened) the objectives and scope were clearly defined, but because of the financial implications (RLG (MH) (M) (1977). As resource constraints in all parts of the system tightened, but the need for action research became more recognised, the question of responsibility for experimental development work appeared on more agenda. The solutions seemed to be *ad hoc,* sometimes drawing on funding by private foundations. Service developments established by field authorities were occasionally taken up for evaluation by RLGs but the problems to which evaluative research was exposed in these circumstances clearly displayed themselves. Local policies and/or organisation of services radically changed, or developments were abandoned or funds withdrawn midway through the research and the conditions under which rigorous study, particularly of a comparative nature, disappeared.

But once again the problems were not purely practical. There are echoes in RLG discussions about evaluative and action research of the conflicts in the scientific community about these subjects. The drive to uphold the scientific criteria of objectivity, rigorous control and generalisability in evaluative social research was matched by doubt about the feasibility of control and generalisations and about the value of objectivity and measurement alone in evaluating human experience. The assumption that evaluative research into practice or service systems was inevitably rendered valueless when the practice or system was changed before the end of the research period was challenged (RLG (MI) FN (81)/1). An RLG in which both 'science'- and 'mission'-oriented advisers were represented was deeply worried about criteria of objectivity and control. But it also paid attention to an experienced researcher complaining about the limitations of an action research project framework which did not allow him to share practitioners' changing experience. In other words, some of the work of RLGs was at the centre of scientific controversies. Scientific advisers were not in a position to give authoritative advice.

Problems of the relationship between development and research and of evaluative and action research methodology were discussed in the CSRC but, as we have seen, to no effect. Evaluative research and methodology were considered by a departmental working group but not further pursued by the CSRC (CSRC (75) M/2; CSRC (76)/19).

At least three RLGs came to ponder on the possible contribution of sponsored research to long-term issues of social and economic structures and policies. Advisers wanted research into social disadvantage and child health, primary prevention, and socioeconomic factors in mental illness. The interests of RLGs at this level had been stimulated by a seminar on social care in 1977, funded by the DHSS and jointly organised with the Centre for Studies in Social Policy (Barnes and Connelly, 1978). RLG chairmen felt that such issues went beyond the remit of individual RLGs and indeed that the work required was probably more appropriate to the research councils. Certainly RLGs were most successful in identifying short- to medium-term research questions within their own policy boundaries, but advisers were less sure that that would or should always be the case (CSRC (78)/1; RLG (MI) (M) 1978, 1979). Again, the issue was given sustained consideration by the CSRC-RLG panel.

Over time, there were some discernible shifts in the balance of RLG activities and in the problems that emerged as significant. Gradually the RLGs spent more time on completed research reports, and on identifying their role in the dissemination of findings and, more controversially, in the policy implications of commissioned research. In 1977 Research Management issued guidelines for the handling by RLGs of final reports of research commissioned by them (RMI (77)/33). Gordon and Meadows' (1981) detailed study of the dissemination practices of the DHSS research management system showed wide variation of practice between RLGs and the differing importance attached to these aspects of the RLG role. Some RLGs considered hardly any final reports, while others regularly discussed them. They varied, too, in their concern with the policy implications, the research implications and the need for active dissemination of completed reports. 'Science-oriented' advisers tended to take the line that it was the researchers' job to disseminate. Other 'mission-oriented' advisers felt that the Department should take a much more active role.

Boundary problems

The boundaries of the RLGs were, we believe, important factors in their effective functioning. But they also created problems. Some seemed, and indeed were, irrational once the nature of the problems encompassed was fully explored. Battered women came under the aegis of the homelessness and addictions RLG, presumably because their impact upon policy-makers derived from their need for housing. Even if homelessness and addictions' remit were broadened to include social pathology as mooted by the group itself (RLG (HA) 78(P)/10), the anomaly noted in the children's RLG that violence in marriage and violence against children in marriages were the province of two separate policy divisions would remain. And rationalisation entailed serious practical problems. It could mean the loss of a policy division, at a time when staffing was already being reduced.

Again, *ad hoc* solutions were found to boundary problems and the RLG panel could identify few clear principles for tackling them (CSRC (78)/1 App. 2). Many policy issues (for example handicapped children, and the elderly mentally infirm) entailed action by more than one policy division. In 1977 all RLGs except one recorded joint activities and joint sponsorships with others (IM (78)/7). More seriously, many issues required co-operation with departments other than the DHSS, particularly the Home Office (forensic psychiatry), the Department of the Environment (homelessness, community care of the elderly) and the Department of Employment (employment of the handicapped and youth unemployment). Cross-departmental machinery for the development of research did not exist. The mechanisms of the intra-departmental joint working group or seminar were used with some effect by RLGs (in particular children's and mental illness RLGs on disturbed children and adolescents) (Joint RLG Sub-Group, 1978). But responsibilities for implementing their work needed to be worked out if they were not to prolong their lives and so complicate the basic RLG structure. Also cross-RLG issues of this kind were vulnerable to any manpower reductions in the Department either in the client group divisions or in the OCS. As new problems emerged for the Department to tackle (for example, the needs of ethnic minority groups) it was likely that existing boundaries would become less closely geared to either policy or research needs. In these circumstances, and in the absence of restructuring, brokerage roles such as those housed in OCS became crucial to the maintenance of systems.

Scientific accountability and the RLGs

Reappraisal of the role of RLGs came with the appointment in 1978 of the first Chief Scientist with an executive role. He determined to increase the scientific strength of the Department's research base. As part of this policy the role of OCS in the commissioning of research was to be more clearly defined and strengthened. Too strong a role for policy-makers through RLGs in the research commissioning process, it was implied, might mean dilution of scientific standards. Additionally, RLG activity was to be significantly reduced. The rationale for the new policies was the imbalance in development of research strategy as between RLG and non-RLG areas of work in the Department, together with the reduced staffing capacity of OCS to service RLGs. This second argument was difficult to sustain in view of the fact that the new arrangements seemed to increase the work of OCS. And the case for reducing commissioning activity by RLGs because more completed work was coming in implied the need for a shift in the activity of RLGs rather than an absolute reduction.

These policies, together with the absence of a role for the RLGs in the revised guidelines for Chief Scientist's visits, were evidence of a decisive attempt to shift power in the Department - perhaps with a view to creating stronger alliances between the Chief Scientist and his advisers. The existing strength of relationship was located in RLGs between scientific advisers and policy-makers.

When the Chief Scientist presented his strategy (IM (79)/10), which in detail would have virtually terminated some RLGs and reduced activities for others, the RLG chairmen strongly resisted it (IM (79)/1 and 22). They stressed the value that they placed on the exchange of views between policy-makers and scientists and asserted that interaction between these groups, OCS staff and service advisers from the field was essential to the fulfilment of RLG functions. They rejected any implied suggestion that the involvement of customers in the commissioning and review of research might undermine its quality and indeed insisted that quality depended on both high scientific standards and policy relevance.

Policy-makers' commitment to collaboration with scientific advisers in RLGs had emerged in a meeting of the children's RLG several months before the Chief Scientist's strategy was formulated (RLG (CH) (M) 1978). The joint working group of the children's and mental illness RLGs on disturbed children and adolescents had reported. Scientific advisers pressed for the continued involvement of both RLGs, particularly their scientific members, in commissioning in this field and in following up a key recommendation of the working party for a new DHSS-funded unit. A paper from the policy division broadly supported their views. They were, however, opposed by the Chief Scientist who considered that now the general strategy and specific research needs of the field had been determined, responsibility for implementation belonged to OCS.

Following the defence of the RLGs by their chairmen, a modified strategy for reduced activity was agreed (CSRC (HPSS) (79)/5). Procedures for new research were to be streamlined. The amount of research handled in most RLGs was to be reduced. As a result, RLGs met less frequently and the volume of research in any case had been reduced. However, there was little evidence of a concomitant reduction in the commitment of policy-makers in RLGs to research or to collaboration with scientific advisers.

Scientific advice in the RLGs

'Government will make use of professionals in so far as they demonstrate consensus about their objectives and/or have a technology to apply their theories to real world phenomena.' (Rose, 1977). A reiterative theme of our book is that scientific advisers to the DHSS were not a homogeneous group. RLGs brought together social and medical scientists, some of whom initially saw their areas of expertise as distinct, but later found that they could collaborate. The experts' views were, however, by no means always unanimous or even complementary, and the differences were within social or medical sciences as well as across the divide. RLGs with different combinations of disciplines appeared to generate different values. The dominant discipline in the early years of the mental illness RLG was psychiatry. During that period values of academic rigour, objectivity and generalisability together with a strong professional perspective on problems predominated. They were challenged when the disciplines represented on the RLG were broadened, in particular with the incorporation of a sociologist. The structural perspectives of sociology are not easily reconcilable with therapeutic perspectives in psychiatry and the distinction between cognitive and value differences is sometimes hard to maintain. Another RLG with advisers including an economist, a specialist in social work and social policy and a medical scientist exhibited very strong concern from the beginning with consumer perspectives and issues of choice. Again, distinct values emerged which seemed to derive from a particular combination of disciplines.

However, there are other ways in which scientific advisers might be differentiated. Advisers can be categorised by an orientation towards either science or mission or power. But the conflicts between and alliances amongst such scientists were not simple. In one RLG 'mission-oriented' reformers and 'science-oriented' advisers were deeply divided. Some members placed most emphasis upon the blurring of boundaries between research and development, research and dissemination. They wanted to fund research that could help practitioners, clients and families. Others were more concerned about validity, bias, the lack of analysis or clear focus and the generalisability of much of the work being funded. The influence of the first group was reflected in the RLG's acceptance that dissemination can and should be built into research and in the focus upon practice and upon families in the second strategy document produced by the group (DHSS Handbook, 1979). The influence of the science-oriented advisers was demonstrated in the failure of the RLG to produce an effective dissemination policy and in a shift of emphasis at one time towards academic and conceptual issues and away from impact and policy. But the two types of advice in the group created alliances too. There were shared judgements about the quality of research and both sets of advisers were united in rejecting too sharp a division between fundamental and applied research.

This RLG, and others too, were faced with the problem of having scientific members who were also major contractors of the RLG, inevitable in a field where work of the highest repute was being funded by the Department. But it meant that there were powerful forces working against major reappraisals of research strategy.

Scientific advisers undoubtedly belonged to interest groups other than the scientific community. Some were quite key figures in the politics of the policy fields concerned (one was Chairman of the National Development Group in his field); some were scientists who knew that their discipline needed a stronger financial and institutional base than the research councils either would or could provide; some were identified with institutions or traditions whose role was under challenge by virtue of central government policies (for example, the shift of responsibility between the National Health Service and local authorities). Such identifications inevitably informed their perspectives and their priorities. They were not always divisible from their scientific experience and judgement.

Exchange relationships and the RLGs

But if the advisers did not emerge as homogeneous, can any analysis be made of the relationships of exchange and power between scientists and the Department at RLG level? We think it can. We have already noted that advisers in some RLGs tended to generate identifiably distinct values, and that, even where there was clear conflict amongst advisers, there were also points of agreement. Moreover, in some RLGs there were, at least for periods, one or more dominant advisers, whose views tended to emerge as representative of the adviser members as a whole.

Our study of the RLGs during the period before the changes of 1979 suggests that RLGs were operating according to at least three different concepts of the relationship between policy-makers and science. The first might be characterised as a 'distance' concept. The mental illness RLG most nearly approximated to this way of working. Scientific advisers were given plenty of space and leadership in the RLG, but the distinctions between science and policy were for the most part kept clear. The scientists wanted to assert influence upon policy, but through ensuring that research commissioned was of the highest possible standard. They attempted to assert stronger scientific control of existing research and applied strict standards to new proposals. Policy-makers, particularly the chairman, kept a low profile, seemingly on the principle that science should provide as reliable an instrument of policy implementation as possible. But dialogue with scientists was not seen as a potential source of policy development. The complementarity between the two sets of views was well sustained in the first three years of the RLG's life. However, when the scientific membership was broadened, the consensus among the scientists weakened and policy-makers began to question the scientists on their own ground - the applicability of strict scientific norms to social policy research. New scientific advisers, for their part, began to assert a stronger role for science in quite fundamental questions of policy.

This RLG thus began to look as if it might move towards the second model of relationship between policy-makers and scientists in which there are more opportunities for interaction and exchange. The children's RLG worked more closely to this model. Its early meetings were marked by more overt adviser challenges to policy-makers and more specific proposals that research should be a source of critique of policy. Policy-makers for their part asserted the administrative and policy framework within which the RLG could work. But in consequence it tackled specific issues in small working parties which made productive dialogue possible between policy-makers and scientific advisers and also amongst scientific advisers of different traditions.

A third pattern in which policy-makers asserted strong leadership was to be found in the RLG for the elderly. The control retained by the policy-makers was tighter and gave scientific advisers little space to make a distinct contribution. It became a highly systematic and initially productive RLG, but there was limited room for the exploration of differences and therefore for a more creative relationship.

Certainly there were different degrees of exchange between scientists and policy-makers across RLGs, but the Department emerges clearly as holding and asserting the greater power. Scientists got access to the policy-makers who, to some extent, opened their minds to them. RLGs served as one of the few neutral forums where policy-makers could consult people outside the system. For some civil servants required to spend their time holding a policy line, RLGs afforded a welcome opportunity to discuss issues in a freer atmosphere. They raised levels of consciousness on some policy issues. Moreover, they helped the DHSS to become more aware of the range of scientific disciplines and research methods available to the study of policy areas. Policy-makers, as we have seen, strongly defended RLGs when they came under attack. Scientists found a forum in which they could question and challenge the Department and a means through which they could strengthen the authority of science in the Department. They were instrumental in increasing the scale of DHSS-funded research so, at least in the short term, strengthening the position of some disciplines and fields of research at the edges of scientific acceptability. But they also helped the Department to control its funded researchers so strengthening its arm to veto as well as to nurture. And while the Department gave advisers room to challenge the limits to their influence upon policy, for example through discussion of policy dissemination, such discussions reached no definite conclusion. The Department for its part got reinforcement against attacks upon its legitimacy through closer collaboration with the scientific community.

The Small Grants Committee

The purpose of the RLGs to harness research to the Department's needs can be contrasted with that of the Small Grants Committee (SGC), specifically established to alleviate researchers' fears that all research monies would go into work invented by the policy-makers. Unique in British government, it embodied the concept of 'free monies' put into the science-government encounter as a concession to science's need to pursue its own ends, or at least the ends of public policy as construed by scientists. It was designed to reach quick decisions in a way readily acceptable to researchers in the field. At the same time it was to divest research management of a considerable work load so that they could concentrate on the needs of policy divisions.

Together with the RLGs, the SGC was a long-term survivor of the post-Rothschild research commissioning system. It began in 1974. It was to receive applications for grants of up to three years of £20 000 (later £40 000), a year, or less, which had not been initiated by the DHSS or by DHSS-funded units. To provide for those occasions when a researcher submitted a proposal which coincidentally fell into a stated priority area of a policy division, the bypass mechanism was used (Korman, 1976). This enabled it to be transferred from the small grants scheme to the policy divisions until the mechanism was abolished in 1980.

The Small Grants Committee's external membership of nine scientists ranging from social work researchers to a pharmacologist reflected its wide remit. It covered all health services and personal social services research, although it excluded research in specialised areas (supplies, equipment, building, social security and computers) which had their own committees. After one year the Chief Scientist handed the chairmanship to an external committee member. Three members of research management representing nursing, social work and medical research were also full committee members. The committee was thus small by departmental standards. It met frequently (bimonthly) and rapidly developed working procedures.

By the beginning of December 1975, some 18 months after the committee was formed, it had considered 112 applications. Thirty-six were transferred to other parts of the Department or to the MRC, and thirteen were withdrawn. Over half were medical, forty-five were concerned with social services, five with nursing and six with NHS organisation (SGC (75)/22). Even the RLGs could not develop the same kind of experience as did the SGC of judging research proposals, of bringing together the views of both medical and social scientists, and of developing a consensus of standards of acceptability. The traffic going through the committee enabled all of this to happen in a way not possible elsewhere within the Chief Scientist's Organisation.

Many of the issues which it tackled were similar to those of the RLGs: criteria of acceptability applied to research proposals, relationships between medical and social scientists, the recruitment and nurturance of new, often inexperienced researchers. Both the RLGs and the SGC were also concerned about the relationship between their two institutions.

The Department proposed that the SGC judge applications on their scientific merit, on relevance to DHSS objectives and priorities, and on the need to identify and encourage researchers. Attempts to match applications against all three criteria proved the committee's most difficult problem. Efforts to set high standards of scientific merit often went against the other two criteria. Questions of policy relevance were the easiest to deal with, as research management, in consultation with policy divisions, would have removed from the small grants scheme those proposals of greatest importance to departmental customers. The encouragement of new workers in underdeveloped fields could mean giving grants to people who had interesting, if uncertain, ideas on the assumption that they would learn best from their own experience. This approach was hard to reconcile with maintaining standards of research. And the objectives of speed and economy of research management manpower conflicted with solutions that entailed protracted negotiations with researchers.

A major concern during its first two years of operation was whether the membership of the committee could handle its full range of applications. This was particularly evident at the early meetings, when social scientists were more reluctant to comment on research involving expertise in medical or health studies than were members with expertise in health services research to comment on social services projects. After two years, however, members felt more confident about working as a whole committee: in fact, it was thought educative to have the two fields together. Similarly, an earlier apprehension that comparison of medical and social services proposals would mean that excessively rigorous criteria would be applied to the latter proved unfounded. Members felt that social services proposals were given a fair hearing. No explicit framework of evaluation was built up, but the common criteria were that the researchers must demonstrate that whatever method they proposed was viable in terms of their own objectives, and that they were sufficiently competent in techniques which their own discipline would endorse. Yet there was a general feeling of disappointment that both social and health service proposals were poorer than expected. In a series of interviews with committee members 18 months after the scheme had started, it was shown that the leading criterion had become that of scientific merit.

This was, indeed, a theme which the SGC discussed fully (SGC (75)/33; SGC(75)/21). One social science member noted that although all applications related to the personal social services were thought to raise important questions, it was disappointing that on scientific grounds so few proved worthy of support.

> Members of the Committee and the professional staff have leaned backwards in their attempt to look at these applications sympathetically, because of the importance of the topics and the desire to encourage individuals or institutions with an interest in research in the personal social services ...wherever possible the Committee has given an application the benefit of the doubt. This appears to some members to introduce the

danger of a double standard, one for bio-medical and another for the social service applications.

Another member of the committee said 'if we wish to support applicants whose proposals, whilst scientifically unsound, contain the germ of a promising idea, we should be more explicit over the reasons for such acceptance and seek if possible an assurance of careful supervision.' This member went on to suggest how the committee and research management could help research in this field to develop. Experienced researchers might act as consultants to promising applicants. The SGC might publicise its purposes and standards and help potential applicants. Research management might promote discussions at the research bases.

A member of research management, responsible for social work research, pointed out its difficulties: the social worker's own self was the main tool in the process; research was in its infancy; often methodologies rested at the first stage of broad concept-building, such as 'action research'; social work was rarely represented as a clinical experiment; social service-social science phenomena were too multivariable to enable the classic research experiment, wherein every factor bar one was controlled, to be anywhere nearly achieved.

The Small Grants Committee demonstrated that a multidisciplinary group of scientists was capable of acting within the tradition of grant-making boards. But it could not, any more than the RLGs, resolve problems of the development of evaluation of pre-paradigmatic research.

Conclusion

We have examined the two longest-lived elements in the DHSS research commissioning system set up with contrasting aims. The RLGs represented a deliberate attempt to steer science, albeit through collaboration with scientists; the Small Grants Committee demonstrated a continuing commitment to scientific initiative, or perhaps a reluctance to let go of the 'golden age' of coincidence between internalist science and policy development.

The two institutions faced some similar issues, but the RLGs' problems were far more complex, bringing together as they did systems with different conceptions of knowledge and different types and degrees of power.

Government's need to reduce complexity pulled against science's tendency to elaborate and to seek for explanations or conceptualisations likely to cut across policy boundaries. Scientists' aspirations to promote pre-paradigmatic science and to encourage multidisciplinary research into relatively unexplored areas of work which needed long timescales conflicted with policy-makers' instrumental approaches to research.

Certain accommodations were reached: science got some nurturance, and scientists as well as policy-makers used boundaries to advantage. Policy-makers' instrumental approaches to research were extended to accommodate the idea that it might help them to develop as well as to implement policy. They learnt too that research could often provide limited answers to problems or none at all.

The relationship between steerage and finalisation of science was underlined. Evaluative research, for example, was a hotbed of scientific conflict. Either problems had to be redefined to be tackled within the limits of existing theories or methods or they had to await new developments. RLGs found themselves faced with such issues of the relationship between policy and research, but they had no conceptual frameworks within which to tackle them. The institutional framework within which they might have been explored, the CSRC, was, as we have seen, disbanded without getting to grips with them.

As the history of the RLGs evolved, conflicts of power interwove with those of epistemology. Many scientists were as keen to influence policy as to develop science. All three RLG 'models' were at bottom models of the containment of scientists by policy-makers. Policy-makers showed appreciation of the distinct nature of research for policy when they asserted the importance of multiple criteria, against the OCS drive from 1978 to confine quality criteria to those of science - but they were also underlining the strength of their alignment with external scientific advisers. Once the Chief Scientist's role became executive, conflicts of power between research management and policy-makers became more evident.

But scientists too displayed hetereogeneity and conflict. Although in both the RLGs and the SGC medical and social scientists found themselves more able to work together than might have been expected, and although some scientific norms were universally held, scientists clashed about their importance relative to each other and to social and political values. Science showed itself to be highly permeable to social and political needs and drives, although this characteristic emerged most strongly when scientists found themselves in the policy arena.

Chapter 8: The Chief Scientist's Organisation and External Research Bases: the Case of the DHSS Research Units*

THE AMBIGUITIES and uncertainties entailed in the Department's attempt to develop policy-related research have constituted a dominant theme of the preceding chapters. The research units funded by the DHSS also experienced an unresolved tension between objectives. Were they to be independent centres of scientific excellence? Was their research to be subordinated and instrumental to policy needs? Were they to collaborate with the Department to define more clearly research relevant to health and personal social services? Different answers to these questions are rooted in different conceptions of science, of its authority and of the relationships between research and policy.

The DHSS inherited two distinct traditions of developing research resources. Two of the departments which contributed to the new DHSS, the Home Office and the Ministry of Social Security, both had strong in-house research capacities, whilst the Ministry of Health sought vigorously to nurture external research, and by the mid-1960s had begun to fund mainly university-based research units.

Between 1973 and 1981 units absorbed most of the funds for research in health and personal social services. In 1979 they were allocated approximately 44 per cent of that budget (Church House meeting, 1979); by 1982 the figure had increased to approximately 65 per cent (ID (82)/1; DHSS, 1982). Units lived through a period of rapid change in the status and power of the university system, in expectations of science and, more particularly, social science.

In the mid-1960s they were seen as a way of helping the emerging fields of social medicine and health services research to establish themselves in the powerful system of medical research, and so to constitute a strong and independent network for the generation of new knowledge on which government could rely. The potential for productive exchange between two powerful systems was thought to be high. By the beginning of the 1980s a quite different picture was emerging: of units heavily

---

* The data base for the material in Chapters 8, 9 and 10 is described in the Appendix to this chapter. A more extensive account can be found in Henkel M. and Kogan M. 1981, *The DHSS Funded Research Units: The Process of Review,* Department of Government, Brunel University.

dependent upon the Department, to which they were as much a burden as a resource.

What is a unit?

The distinction between units and programmes long remained unclear. Although units were established well before the Rothschild Report, they were not differentiated from programmes in the DHSS *Handbook on Research and Development.* By 1980 the Department had defined a unit as a programme on, or about to be put on, a rolling contract. But in 1991, eleven designated units out of 34 were still on fixed-term contracts (DHSS, 1982 ibid).

At one time the Department probably hoped to sponsor entities strong enough to compare with the best of the MRC units. They would build a network of policy-related science with its own peer group and able to train good people to do useful work. The principles underpinning MRC units were described by the Secretary of the MRC in his evidence to the PAC in 1979 (PAC, 14 March 1979, 1288, 1292, 1293). They were built around a good man (sic), who was given tenure - as were many of the research staff - for science developed best when good scientists were allowed to follow their own ideas. Projects were part of a seamless garment, not easily separable elements in a total unit programme (ibid).

DHSS units, for their part, were described by a research manager in 1977 as *'sui generis'* (IM (77)/1*).* Their constitution was 'a compromise and a bargain' designed to achieve the maximum independence, objectivity, intellectual standing and strength that could be combined in one institution or group of people with responsiveness, understanding, involvement with the problems of the Department (and, the author might have added, total or substantial financial dependence on it).

Another tension was noted. The Department had, through the Chief Scientist, looked for 'scientific originality in every team', but by implication also expected all units to share in the 'relatively humdrum work' (ibid, 1977). There was, in fact, a DHSS context within which the units were to work, a context of policy, decision-making and practice in the field, to be put alongside the determination to develop good science.

Contracts were made round named directors and subject to re-negotiation if they left their institutions. Thus the DHSS, like the MRC, asserted its faith in individual leadership. But the commitment of the Department to a unit director was different from that of the MRC. Contracts normally rolled for six years, with biennial review. Some continuity was thus established but not the full security of tenured academic appointment, though some directors had tenure from their home institutions. Usually the director was provided with a 'basic core' of staff, funded for the life of the unit, to promote general research capability. Other staff were appointed for specific projects.

The field of work was identified in the contract. Its breadth varied greatly, and units' terms of reference sometimes extended beyond research. Two units in our group of five (and one of the further five which we have studied) included dissemination, and one training, in their remit.

In the post-Rothschild commissioning system the Department attempted to structure units' work around specific projects. This was a slow process - by 1981 only 50 per cent of rolling contracts' work was devoted to customer-requested research (DHSS, 1982) - but the expectation that units would accept commissions was made clear as in the correspondence on one file, 'promises of long term support ... can only be made on the understanding that the unit undertakes tasks commissioned for the Department when we believe that the research we want done ... is more important than the unit's own proposals.' Following this, a condition was written into contracts that commissions 'will not unreasonably be declined if they are scientifically acceptable, financially viable and can be carried out within the agreed resources of the unit.'

Attempts were made to mitigate the constraints upon researchers' freedom by the adoption of Rothschild's proposals that up to 10 per cent of contractors' time could be devoted to 'General Research', that is, research not directly concerned with the commissioned programme (Rothschild, 1971, para 19). But again practice varied. One unit had correspondence to show that this percentage was not fixed and could be considerably extended. Another, the most recently established in our study, had no degree of freedom incorporated in its contract. The idea that units might be 'centres of excellence' had not, however, disappeared from the Department's thinking. Another contract referred to a unit as a high quality research resource which would contribute to the general development of its field - a clear indication that this unit was to be such a centre.

Contracts contained an explicit requirement to observe the ethics of confidentiality and the protection of subjects of research, and units were required to submit proposed surveys of service users to the scrutiny of the Office of Population Censuses and Surveys (OPCS). Correspondence with one unit in 1977 suggested that this was as much for the protection of the public as for the assurance of science. On the other hand, the Department's publication policy placed few restrictions on researchers. Accountability to the Department was provided for through annual unit reports, reports of completed work, and the quadrennial visit of the Chief Scientist. Communication with the Department was systematised through the liaison officer.

Tension between freedom and control, and between centres of excellence creating their own coherence and centres that can respond flexibly to government needs, was, then, a pervasive feature of unit contracts. The balance between these values shifted and they emerged with differing force at different times.

A brief history of DHSS-funded units

Some of the units go back to the 1950s but it was in the 1960s that officials from the then Ministry of Health sought out and funded researchers in social medicine and health services research. The first were established as such in 1965 (DHSS, 1982). The Department's research management policy went, as we have suggested, through three distinct phases each embodying principles, the conflicts between which have had a direct impact on research units.

The first was an individualist, perhaps charismatic, phase. Scientists were discovered and units promoted by individual members of the Department, whose images are now more heroic than bureaucratic. The 'intemalist' conception of science was dominant: if good researchers were given freedom they would produce good research that would advance the development of health services. Unit directors whom we found to be nostalgic for this period remember it as one of receptiveness and encouragement by the Department. Relationships were informal and mainly with a higher level in the Department than now. Unit directors felt valued and, to an extent, influential. The ethos was one of equality and reciprocity in which issues of independence and control did not obtrude, although there were Department-initiated commissions.

The implementation by the DHSS of the Rothschild Report thus had large implications for the units, although its impact was gradual. For example, the customer-contractor principle did not in itself provide an immediate threat to researchers; some hardly noticed its arrival, others actively welcomed it while observing little change in practice. This crept upon units slowly. One director spoke of projects becoming more tightly defined and time specific, while another found the balance between fundamental and applied research in his unit shifting (IR(80)/1; IR (80)/6). Economies affected the shape of the research budget and the costs and benefits of units' work assumed a greater importance for research management. The units had been the cornerstone of the Department's policy for health services research. Now the Department shifted from an individualistic to a systematic approach. Committees were established to make and coordinate research policy. The units were particularly affected by the research liaison groups set up to promote neglected research, the categories of which fitted uneasily with the work of the units.

For some unit directors the impact of these developments was lessened by their recruitment to RLGs or other committees. But by 1975 the units were perceived as posing problems of administration, resource use and function. Many units were working on areas without RLGs, and some spanned the boundaries of several policy divisions (up to nine it was thought) (IM(75)/1). Liaison with them consumed, to some, an unacceptably high proportion of research management time. They were the most conspicuous resource that the Department possessed, but at a time when research talent was thought scarce, it was being diffused and directed largely away from the Department's defined priorities. There was a strong lobby in research management to bring more research units into relationship with RLGs, if possible through specialisation in one RLG area.

In 1977 a seminar organised by the DHSS with the directors of twelve health service research units discussed relationships with RLGs. In 1977/78, all the RLGs except those for the elderly and reproduction and allied services had at least one specialist unit for its field, and all except forensic psychiatry were funding work by multiple units. The RLG for physical disablement was spending as much on research in five multiple units as in three specialist units. The Department's hopes of outcomes from the seminar were by no means unanimous, perhaps because the issues were far-reaching: the freedom to be accorded to units; whether their role should shift towards the instrumental and reactive and away from that of developmental 'centres of excellence'; and indeed whether there was necessarily a conflict between those two (ibid,1979).

The issues remained unresolved. Health services research units maintained substantial areas of work outside RLG boundaries. Unit directors wanted to keep direct links with policy divisions through which they could monitor the continuing relevance of their work to the Department and its impact. But such relationships were not easily incorporated into a unified research policy and management system.

By 1979 units' capacity to give value for money in the quality of their science was as much questioned as their contribution to the Department's priorities. For some, the issue had now become one of survival; all were soon to be caught in the cross-fire of attacks upon the civil service and the higher education sector.

Health and personal social services research

In considering how units' work affected institutional relationships, and in particular those with the DHSS, we shall focus upon issues of coherence and of power. The reiterated theme is the influence upon the cognitive structures, and intellectual definition of their field of work, of social and political forces.

Health and personal social services (HPSS) research encompassed fields still evolving from different starting-points. Their substantial DHSS sponsorship was perhaps their most powerful defining force.

Originally the Department's units were in health services research, predominantly founded upon epidemiology but medical sociology and information science were also represented. Health services research was now institutionalised, not least in the concordat reached between the DHSS and the MRC in 1980 (DHSS, 1980). Health services research groups and health economists had been particularly active in defining it (Aberdeen, 1980; Williams, 1981).

But a definition of the field remained elusive. Taylor (1981) places health services research at one end of a spectrum, the centre of which is clinical research and the other end, biomedical research. The fear identified by Taylor was that: 'attempts to extend medical authority out from the central [clinical] area of [health research] where that profession's competence is most relevant may be a potent source of discontent.' (ibid). Health services research is a point where medicine intersects with social science disciplines and a field in which 'health concepts merge with those of social and economic distress' and also deviance and social control. The boundaries round 'social care' or personal social services research were still harder to define (Nodder, 1979).

A modified form of Wing's categorisation of the research of interest to planners (Barnes and Connelly, 1978) provides a framework within which to describe HPSS research. It was concerned with the nature and causes of disablement, and of social and behavioural problems. It entailed theories and methods of care, prevention and treatment of these problems. It involved studies of the systems through which care, prevention and treatment are mediated and of their theoretical, organisational and policy problems. Often the researchers sought to analyse scientific method in fields of study such as general medical practice, nursing care and social work, which had only emerged in the last 25 years. There was a wide range of tasks: describing, defining, measuring and determining the incidence and prevalence of all sorts of disabling conditions and deviant behaviour. Handicap, mental illness, addictions, family breakdown and family violence afforded a few examples. Researchers were tracking the natural history of diseases; but they also generated hypotheses and theories of correlation and causation to explain and predict these phenomena.

There was work to be done in mapping, analysing and comparing existing ways to tackle problems and the institutions within which work is carried out. New theories and methods of intervention, and new hypotheses explaining how and why institutions function, were generated. Attempts were made to establish criteria and methods of evaluation.

Almost every DHSS unit in health and personal social services research spanned at least two of the categories or levels of work described, and some moved between all three.

Theoretical conflicts

Major disciplines involved included epidemiology, sociology, economics, psychology, anthropology, psychiatry, statistics and social policy. The work of the older units reflected shifts in thinking over the last twenty years about the relationship of the natural and social sciences, and about the balance between analysis and synthesis, linear and systems thinking, and objective and subjective knowledge of social life. An influential researcher in social work has emphasised the value of an epidemiological approach to social services research (Goldberg and Warburton, 1979). But the movement against positivism in social science was now riding high. The clearest contrasts were between those in the units who took a positivist/analytic approach and those who were interested in a systems/development approach.

From their own accounts several units had built up their work through carefully negotiated access to the field and meticulous description and measurement. The development of reliable instruments for calculating prevalence of conditions often took years of work. Similarly, the refinement of research design in investigating causal relationships between environment and health, between social factors and behaviour, and between different forms of intervention and outcome in health and social care had been long and rigorous. The aim was to develop the tightest possible controls that the investigation of social and environmental conditions could allow, in the knowledge that the quality of science achieved would never be as 'good' as that found in the best laboratory-based science (IR (80)/23) - the key criteria of merit here being precision and reliability. It was important in this view that science was exploited in the understanding of this field of work, but equally important that it knew the limits of its competence. This was an approach founded in analytic, discipline-based research.

Units interested in anthropological, interactionist and ecological frameworks did not indiscriminately abandon objectivity or control; but they criticised positivist values of science. Some emphasised that health and personal social services research was directed to finding ways through human social problems arising from rapid change. Trist (1972) argues that such 'domain based research' requires an integrative scientific strategy in which representatives of different disciplines come together, not simply to aggregate their knowledge, but to use it creatively as a group. Moreover, it requires collaboration across the boundaries of science, policy and profession.

Some research units seemed to find the analytic, discipline-based model of research not fully operable, and they took on some of the features of the second approach without necessarily embracing its basic, and radical, assumptions.

Some researchers whose work originated in disciplines such as epidemiology and psychiatry in the study of mental illness in the community extended their working systems. Some increased the number of disciplines encompassed, some had practitioners as research associates, some did both. One unit with nine core staff included six disciplines; another had among its project staff a health visitor, a manager of an adult training centre, two teachers and a social worker. The unit studying mental illness in the community incorporated the study of social work into what was originally a medical field. Another unit found that in five years its focus shifted from disabled children and their families to the far wider concept of family policy. Finally, a unit studying the mentally handicapped extended its interest into the development of normal children and from the individual to the social system of the handicapped person. Almost all these changes derived from developments in the units' thinking about their work and not from pressure from the Department. Their work seems to have led them towards synthesis, complexity and elaboration as against analysis and simplicity of formulation. This observation has large implications for understanding what are appropriate scientific methods and intra- and inter-institutional relationships in such work.

Intra-unit relationships

It is generally assumed that health and personal social services research is multidisciplinary. Multidisciplinary work calls 'not only for a particularly agile intellectual make up, but also for a patient and tolerant temperament and for certain social skills (Aberdeen, ibid)'. A team bringing together an economist, a sociologist, a social administrator, a community medicine specialist and statistician to study alternative patterns of care for the handicapped would have barriers of language, modes of analysis, methodology and problem formulation to surmount. The intellectual challenge is hard, a fact that unit structures and contracts did not always take into account. Academic disciplines do not stand still while they are applied. Academics in multidisciplinary settings were subject to threeway pulls - to contribute from existing knowledge, to advance that knowledge, and to comprehend the perspectives of their colleagues (Buxton, 1981). But membership of a multidisciplinary unit could sometimes mean, for an academic trained in an established discipline, a significant shift of identity. Networks for consultation, critique and publication might change, shrink or even disappear. Finally, there might be value conflicts amongst colleagues between the validity of subjective feeling as against objective measurement, of consumer as against professional judgement, of the criteria of cost effectiveness as against acceptability, or of health as against welfare.

Units in our sample dealt with such problems in a number of ways which can be related to their positions on the positivist-analytic, systems-development dimensions of research. One discipline might remain dominant through its pairing with different disciplines and through ensuring that, in a medically-directed unit, for example, orientations of the social scientists were not likely to conflict with those of their colleagues. A unit emphasising the quantifiable measures of health would ensure that its social scientists were economists, statisticians or quantitatively-oriented sociologists. An alternative was to ensure that workers whose perspectives were most likely to conflict - for example, psychiatrists in the chemotherapeutic tradition and interactionist sociologists - worked on different projects. These approaches were different from one of 'equal partnership' in which, say, epidemiologists and economists evolved joint methodology and joint objectives in an evaluative study of patterns of care for the mentally ill. This in turn can be distinguished from a matrix model in which the disciplinary identity of individual workers had been subsumed in a problem to be solved (care of the confused elderly) or in a field in which disciplinary boundaries are blurred (social work). Our empirical work with funded units provided examples of each of these forms of multidisciplinarity, entailing different kinds of organisation and relationship.

Intra-unit relationships were complicated by professional as well as disciplinary origins. Of the ten units (and one sub-unit) in our sample, three employed members of professions (nurses, a physiotherapist and social workers) as research staff. In two research units GP researchers continued their medical practice, while the unit most committed to action research employed a teacher, a social worker, a health visitor and the manager of a training centre whose primary role was not research; but all were clearly seen as integral members of research teams. This unit was working towards a matrix model of research teams. Of the remaining units, one included development in all its projects, and another was engaged in setting up a demonstration project. Both were thus involved in collaborative work with practitioners. Twenty years later these various units would be understood as exemplifying different forms of Mode 2 production of knowledge.

Research unit networks

Of the unit directors in our sample, a minority were predominantly science oriented. The remainder, although regarding themselves as members of the scientific community, were primarily policy oriented. They wanted to affect policies, services, professional practice and the lives of clients. The field was not only their laboratory but their *raison d'etre.* When asked about their audiences, they saw the DHSS, the centre of the policy system, as their prime audience, although many were frustrated by its many faces and by its secondary customer status. Several felt thwarted because their findings might not reach, or change the perceptions of, policy-makers and practitioners. They wanted more access to the policy-makers, and four had sought out opportunities for involvement in the politics of their field. Five identified health and/or social services authorities as audiences, and all saw professional conferences and journals as part of their network. Six rejected notions of clear boundaries between research and development, education, dissemination, policy and practice.

Three directors saw themselves primarily as scholars. One of these thought confinement to applied research to be 'death to the scientist' (IRC (80)/6). His international scientific network was most important. Another director in the same field, epidemiology, similarly believed that it required an international perspective. But two spoke of scientific audiences as important primarily for the careers of their staff, while another identified his field as having, as yet, little scientific coherence. Moreover, his programme allowed little time for any comprehensive reflection on the literature, or theoretical development for practice. About half of the work of the ten units was essentially pre-paradigmatic. The most coherent of the units, in terms of discipline base and field of application, revealed a huge variety of outlets from *Proceedings of the Royal Statistical Society,* to *Social Work Today.*

In short, the units had multiple and shifting networks and their staffs had different audiences and dominant influences. The concerns of directors might differ from those of their staff, particularly younger, unestablished researchers. Directors had increasingly to keep their eyes upon sources of funding and support. This might mean cutting off their researchers from lines of inquiry to which they were intellectually committed and which might enhance their disciplinary base.

For directors there might be tension between the norms of the scientific world and those of policy which must be held in the interests of the staff and of the unit too if the quality of work was to be maintained.

Coherence and power

The research units did not therefore belong to a coherent scientific community and their power differed according to scientific bases. The medical research tradition is that fundamental and applied research may be theoretically distinct but that they should not be organisationally separated; continuity of work between the two is essential (MRC Evidence to PAC, op. cit). Researchers are often members of the medical profession, a system in which research, education and practice are strongly connected. These characteristics give them power.

Medically-directed units contrasted with those directed by members of other professions, nursing and social work. Neither of these had high status or a tradition of strong linkage between research, education and practice. However, they differed in normative and institutional coherence, nursing standing high on both of these counts. This helped nursing to commit itself to establishing a research base, and their researchers benefited from that.

The research-profession axis is one source of understanding about power. Another is the institutional base of research units. In the year 1977-8 only seven of the DHSS-funded research units were not part of either a prestigious hospital or a university, and of those one was in a polytechnic and another was directed by a university professor on a part-time basis (DHSS Handbook, op. cit). Of the ten units in our sample two were not university-based, but of the remaining eight, three had directors without tenure. Other factors, too, suggest that the universities do not necessarily provide a strong support system for research institutions. One of the present authors has commented elsewhere that 'if they are successful, [research units] may earn the relevant disciplinary group and the parent institution as a whole much credit; but they do not thereby acquire greater political power. (Becher and Kogan, 1980, 1992). Two unit directors indicated that their teaching capacity was as strong a unit of exchange in their quest for recognition in the university as their research activity. A third unit, medically-directed, stated the provision of a strong resource for undergraduate teaching to be a key objective. Clinical medicine was still the most prestigious activity in the profession. The establishment of, in this case, a research unit in epidemiology that could play a strong role in medical education could be seen as a means of modifying the traditionally low position of a particular discipline in the medical profession.

Although epidemiology was a dominant academic discipline in units funded for health services research, it lay towards the edge of medical science and was becoming increasingly dependent upon the social sciences. It thus rested on the boundaries of established academic tradition. A linked problem existed for all research units based in universities. They were almost always multidisciplinary. University structures, reward systems and audiences still leaned towards unidisciplinarity (Trist op. cit, 1972). Units might therefore cut across the host institution's decision-making structures and so be in a weak position in their internal politics.

Units concerned with the health and personal social services might develop strong ties with them. Others whose orientation was also towards client groups might be drawn towards a service role particularly if clients were under-provided for. Units might include pressure groups as important members of their networks. Where policy and practice issues constituted units' *raison d'etre* they might exert more influence and provide more recognition than did the home institution or the academic peer group. Research into problems of professionals, bureaucratic organisations, government departments or families might promote sympathy with norms conflicting with those dominant in the university, particularly if such work entailed substantial periods of time away from the unit. One unit was picketed by students critical of its connection with a government department. Only when it increased its teaching activity did the students' attitudes change (IR (80)/33).

The DHSS prized units' connections with universities. But units had to face so many ways and this left them vulnerable and increasingly dependent upon the DHSS.

In the face of the demands of multidisciplinary work, time and security were felt by many researchers to be crucial. The enterprise might be abortive or prematurely terminated if provision was not made for the build-up of trust and for learning new assumptions. In 1979 the DHSS responded to increasing pressure for greater security of tenure, at least for core unit staff. Units who joined their scheme were to offer vacancies first to researchers from other funded units from which DHSS support was withdrawn or reduced (Church House meeting, op. cit). Such measures could hardly be effective at a time when support was being withdrawn from a number of units.

Although previous studies have established some of the problems described to be chronic in research units at large (Platt, 1976) some were particular to the DHSS units. Others sharpened in the economic and political context.

As our direct observation of the units came to an end in early 1981, an increasing proportion of DHSS research funds was being devoted to maintaining a reduced number of units. More than ever the Department needed clear criteria for their allocation. Less than ever, in the face of the complexity described, was it likely to find them. We go on to give an account of the DHSS's attempt to establish criteria based

Appendix:

Note on field work conducted for study of DHSS-funded research units

The evidence for our study of the units, and their review, was gained mainly between 1978 and April 1981. We first interviewed twelve unit directors in 1978. In 1980 the Chief Scientist invited us to observe visits to five research units under the revised arrangements for their evaluation, and to then follow the process of their review by customers.

The units studied were the subject of Chief Scientist's visits within the first seven months of 1980. The DHSS gave us access to their files and to papers on the review procedures prepared by the Office of the Chief Scientist. We saw documents prepared by the Department and units for the visits, the reports of the visits themselves, the comments of the unit directors on the reports, and the reports of subsequent discussions and decisions in the Department. We interviewed five groups of people concerned with these visits. The research units' liaison officers in the Department and the unit directors were seen before and after the visits. In all units except one we interviewed all of the researchers with the lead responsibility for, or a close connection with, the presentations of work made for the Chief Scientist's visits and also saw some of the other research staff. We interviewed twelve scientific advisers from the visiting panels, including the three chairmen who substituted for the Chief Scientist, and two other advisers for each visit. One of these participated twice. We interviewed customers of all the units individually or in groups and we attended three RLG meetings at which the reports of the Chief Scientist's visits to units were discussed.

We also interviewed the directors of five other units not currently under review and were given access to their files. This extended our understanding of the working of research units, their relationships with the DHSS, and the range of work undertaken by units.

Interviews conducted for the unit study are shown in Table 8. I.

We jointly conducted 34 of the total of 75 interviews and, except in two cases, both of us took notes. One of us wrote up the interview which was then checked, corrected and in some cases amplified by the other. Many of the interviews conducted singly were of individuals connected with the work of others jointly interviewed.

Table 8.1 The review of five units:
Interviews conducted between January 1980 and April 1981

	Numbers interviewed	
Roles interviewed	Before Chief *Scientist's visit*	After Chief *Scientist's visit*
Liaison officers	7[1]	5
Policy-makers, DHSS	1	14
Unit directors	5	5
Research unit staff	-	26[2]
Scientific advisers	-	12
Total	13	62

[1] This figure includes two people who had moved from an LO post after planning for a Chief Scientist's visit to the unit concerned had been initiated.
[2] This figure includes the administrator of one unit.

Chapter 9: Review of Units and Scientific Merit: Chief Scientist's Visits*

IN 1979 the Office of the Chief Scientist (OCS) set out to clarify and make more rigorous the quadrennial review of the Department's funded units.

Issues of authority, power and dependence are inevitably heightened at the point of review, but our study was conducted at a time of peculiarly acute tension. The DHSS was reappraising its commissioned research in a climate of contraction and disengagement. Government's needs and its determination to review had become tighter.

The Chief Scientist wanted to enhance scientific accountability and the change in his role from advisory to executive was to improve the quality of research and thus strengthen the authority of science in the Department. Decisions about units had normally been taken in the light of their own history; review must now make a more systematic contribution to setting priorities in research policy. It was also thought important to ensure that units were giving the Department scientific value for money. The difficulties of defining what that meant became evident: the range of work subsumed under the heading of health and personal social services research, the problem of setting firm boundaries between research, service, dissemination and education and the ambiguities inherent in the concept of a DHSS-funded research unit.

The OCS's starting-point in revising the review procedures was that scientific merit should be assessed separately from policy relevance (IM (79)/23). Scientific merit was determined by the quadrennial visit to the units by the Chief Scientist and his scientific advisers. Under the new arrangements, for fear that scientific merit might be confused with policy or practice usefulness, the visiting party normally no longer included representatives from the policy divisions. The Department conducted a separate review of the customers' needs for a unit's work once the report of the Chief Scientist on the quality of its science was cleared.

---

* The monograph on which this chapter is based was cleared in draft by all ten unit directors included in our study and by most of their DHSS liaison officers.

Our study

Our study of the new procedures focused on the Chief Scientist's visit to five units in the first half of 1980. We also compared these with five other units not being visited. The zones of inquiry of the visited units were nursing education, mental illness in the community (identified mainly in general practice and in local authority social services departments), the psychological and educational factors affecting the development of the mentally handicapped, general practitioner behaviour and problems that need to be investigated in the context of general practice, and health services research. A wide range of stages reached and of scientific assumptions were all present in 'our' units.

Arrangements for the Chief Scientist's visit

Units were normally given at least six months' notice of a visit and a firm date fixed at least three months ahead. Units submitted papers for the visit two months in advance and these, together with information from the DHSS about the purpose and format of the day, were to be distributed to the visitors one month before the day itself. After the visit, the report had to be confirmed by the Chief Scientist and his advisers, sent in draft to the unit director for factual vetting, and then in its final version submitted to the director for comments on its substance. This process was to be completed within three months of the visit so that the review with the customers could then begin. In practice, these time limits were sometimes found unacceptably tight and were not always kept.

DHSS preparation (ID (80)/8) critically affected a unit's performance and was compared by a member of the Department to the period before the starting gun is fired in a yacht race where the competitors are manoeuvring for the most advantageous position. The Department's liaison officers knew the work of the unit better than anyone else in the OCS and their advice to the Chief Scientist on the selection of the visiting party was extremely important. This selection was considered by interviewees of all parties to the visits to be potentially the single most important influence upon and source of dissatisfaction about the process.

Liaison officers also noted the importance of their advice to units on the documentation required, on the choice of projects for detailed presentation, and on the structure of the day's programme, although on all these issues the final choice lay with the director. While most liaison officers hoped to help the units to present themselves as well as possible, the judgements made by the directors about how to manage the visit were 'part of the test'. No liaison officer felt they should act as advocates for their units in the Department. Visits were normally chaired by the Chief Scientist accompanied by the unit's liaison officer, who advised the visitors on details of the contract and on the background to the unit's work, and compiled the report on the visit. If the Chief Scientist was unable to chair the meeting an experienced scientific adviser was asked to take his place.

Selection of the visiting party was made through consultation between the unit's liaison officer and the Chief Scientist. Directors could comment upon its membership once the list was decided, but they had no power of veto or of appointing 'internal' advisers as was the case with MRC visits.

Visiting parties did not necessarily mirror the unit's specialist interests. Specialist knowledge could be supplemented through external referees. Scientific advisers from RLGs interested in the work of the unit might be selected if their field of knowledge was appropriate, but normally at least one member of visiting parties was a scientist from outside the unit's field of work and thus tapped into the evaluative criteria asserted over the whole range of research units. This role was often carried by the chairman.

In two visits out of five, the Chief Scientist was in the Chair and much of the discussion was structured by the questions and preoccupations that he brought to the visit. Other chairmen played more coordinative roles, since if the Chief Scientist was absent, all members of the visiting party were advisory only. He, however, carried departmental executive responsibility for the decisions made about units. One visiting party, not chaired by the Chief Scientist, seemed uncomfortably aware of its advisory status. Members became anxious about how the Department would interpret their reported conclusions. They feared that unduly harsh action might result from their criticisms. The presence of the Chief Scientist might have reinforced their understanding of the practical implications of the event from the beginning.

The variation in size of the visiting party was not wholly explicable by the size and range of interests of the units. The units which felt there to be most problems of match between the visitors' expertise and their own concerns were those containing the widest span of interests or disciplines. In one case the Department grasped the nettle by inviting visitors combining expertise in their own fields with acknowledged breadth of experience beyond them. There were other problems of selection. All the units except one had been visited previously. In three cases continuity was sustained in the choice of the visiting parties. But as the team received no documentation about the judgement of the previous visit, advisers who visited twice had to balance the advantage of continuity of judgement against the possibility of incorporating bias.

The units working primarily in the field of one RLG had another means of incorporating continuity, although one RLG member experienced conflict between his RLG membership and his visitor role (IE (81)/1). He found it difficult to confine his interests to the issue of scientific merit and to dampen down his interest in relevance.

Having RLG members in the party was not identified as a source of bias. Bias was ascribed rather to the close knit or controversial nature of many specialist scientific networks and to the strong commitments that might divide different schools of thought (IE (80)/3). This is known to be a difficulty even in well-established fields of science. *A fortiori* it is a problem in health and personal social services research where the controversies about the nature of validity or generalisability often go to the root of the work. In practice, exercise of bias is not easy to differentiate from the application of differences in perspective or basic premises.

The unit's written presentation of its work for the visit was made within a framework laid down by the Department (IM (79)/23). Directors were expected to give an overview of their total research programme, to outline research in progress or recently completed for the DHSS, and set out the objectives, design, methodology and, where appropriate, the conclusions of each project. The director also included a forward look placing the unit's work in the context of research in that policy field and outlined proposals for future work. Most research unit staff interviewed felt that a great deal of time had been invested in producing material for the visit. Some found it helpful to have to take stock and to summarise their work for external scrutiny. Others felt that it was an interruption to their schedule and produced a 'poor return'. The main variable here was the stage reached. For some, the visit came too late to be a useful aid to reappraisal of method or theoretical assumptions, and too early to form part of the writing-up process.

Scientific advisers on the visiting teams received all the unit documentation but most of them were asked individually, or as one of a pair, to lead the examination of one area of work. Some thought that there should be less material compiled specifically for the occasion. Instead, advisers' attention should be drawn to the unit's published work, particularly that considered of most value by the unit itself. But while this might strengthen knowledge of the unit's proven scientific capability, it might not show clearly enough the quality of the work undertaken within the unit's mandate from the DHSS. It reflected, too, the potential that the DHSS might tap rather than the actual contribution made, so broadening the purpose of the Chief Scientist's visit from scientific accountability. Whether that would be desirable is a separate issue.

Visiting parties were briefed by the Department; they received papers outlining the purposes and timing of Chief Scientist's visits, but to no set formula. The briefing in all cases made it clear that the visit was concerned with the scientific quality of the research being carried out by the unit although one briefing also incorporated relevance to a departmental strategy statement.

The background material about the unit's field of work, history and relationship with the Department varied, and advisers whom we interviewed differed in their briefing needs. All felt clear enough about the purpose of the exercise and four were quite satisfied with their briefing. Various kinds of contextual information were thought important: about the use to which particular pieces of research would be put and about individual units' histories, outputs, aims and relationships with the Department. Most who mentioned these thought them essential for the making of a judgement of quality. There was then more general information about the nature and purpose of research for government and how it is managed. One adviser, however, also drew attention to the complexity of the evaluation process and to skills other than scientific required to undertake it. He would have liked the DHSS to organise a seminar to discuss issues of this kind.

Responsibility for planning the visit's programme was shared between the Department and unit director. The setting, selection of work for examination on the day, mode of presentation and style of hospitality were chosen by the unit; while the essential components were set by the Department. These were: a short preliminary private meeting of the visiting panel; encounters between researchers and visitors that allowed for pursuit of questions raised for the panel by reading the documents; a second private meeting of the panel to discuss their judgements; a final meeting with the director and senior staff; and, not least, regular refreshment.

Some units in our study arranged for the visits to take place in the unit's own accommodation, while others held them in rooms designated by the main institution for formal occasions. Some contrasts in styles of hospitality emerged between, for example, emphasis on the work ethic and informality in one unit, and upon efficiency, formal organisation and the place of the unit in the larger institution in another. In all cases lunch was an important part of the day. In particular, senior staff from the units used this time to convey to the visitors more about the philosophy, management and politics of the unit than could be accommodated in the formal programme. And the visiting party were able to get a more rounded impression of the unit's researchers and their work.

The length of the visit varied little and took no apparent account of the differences between different units' size and range of work. The day was dominated by the clock and many participants were oppressed by a sense of rush. Nor were there arrangements for the visitors to meet together for systematic discussion the night before the event. Advisers critical of the arrangements thought that they needed to allocate tasks between themselves and to know whether their judgements on the written material were complementary. They could then devise a strategy to deal with differences and try to ensure optimum coverage in the time available. They could also acquaint themselves with other team members and so prevent a build-up of interpersonal forces that might undermine objective judgement. Two participants in our study spoke of the desirability of getting any 'peacock' display between advisers over and done with before the events of the day itself.

The day of the Chief Scientist's visit

A Chief Scientist's visit was an unequivocally formal occasion. Dress was by near universal consent quasi-*'sub-fusc'*, saving the odd red tie and tweed or corduroy suit. Exchanges between the visitors and the research unit staff were almost without exception characterised by a measured politeness. Only the use of first or surnames differentiated between established and new relationships. And even the most aggressive examinations emerged as elements of a ritual comprising distinctive stages and components.

In private meetings of the visiting party, the chairman reiterated the purpose of the visit - that it was confined to the assessment of scientific merit and excluded questions of relevance to policy. In visits led by the Chief Scientist himself he elaborated his insistence upon the separation by stating that the focus was on the *solution* being pursued by the researchers. The *questions* were of interest only in so far as the kinds of question which the researchers tackled might indicate their scientific capability. Science was not to be judged as dependent upon the nature of the question addressed. The same insistence upon the independence of science was discernible in the answer given to the question whether the visiting party was looking for science of a practical rather than a theoretical kind. What was sought 'was good science which can be applied.' (IR (80)/17).

One adviser suggested that scientific merit was not an absolute but was dependent upon resources available. It was then conceded that the degree of certainty demanded of scientific work partly depended upon the resources that could be allocated to a project. In another visit, where the resources allocated to the unit were large, cost effectiveness was thought to be an unavoidable element in the judgements to be made by the scientists as well as by the policy-makers.

In introductory statements to the visiting party, three directors indicated the importance of integrating research with practice and service development. One made plain his belief that the prime scientific value of his unit's work was in the fundamental rather than the applied programme. A large framework was needed for intensive work and DHSS funding for extensive research was not available. By implication, the scientific merit of his unit could not be assessed by reference only to research undertaken for the DHSS. In different ways these directors were questioning the assumptions behind the Chief Scientist's visit: some were challenging the isolation of scientific merit, others the fragmentation of material presented.

The 'art form' of interchange between researchers and the visiting party was that of a *viva voce* examination. The exercise was strictly time limited and varied by only half an hour. But differences in the amount of work covered were huge. One unit attempted to expose, in presentations made in 2 1/2 hours, the essence of a multiple research programme based on seventeen years' work. At the other extreme, a recently established unit presented within a similar time period all three of its projects. A third unit allowed 10 minutes for the discussion of each group of presentations, while a fourth scheduled 35 minutes for questions on each project team's work. Researchers' expectations of these encounters ranged from them being an essential but tedious exercise to a developmental exchange between peers. Units varied, too, in style of presentation. Two chose the lecture-demonstration in which, in the hands of the most confident exponents, the performance element of the occasion was underlined. Two others concentrated on producing summaries for the team, while one unit made no formal introduction of the work on the day preferring to leave all the time available for questioning.

An adversarial and inquisitorial mode of encounter in which scholars are challenged to defend their work is well established in the academic community. Weaknesses or uncertainties are exposed and theoretical bases or methodology may be held up to scrutiny against alternative approaches. This was the predominant mode adopted in visits observed, although some advisers were more persistent and confronting than others. In three visits one adviser adopted, or moved in and out of, a contrasting 'interactive' style in which he or she displayed personal interest in the work and invited some sharing of the problems entailed. But this was not the expected mode. In one visit the chairman explicitly warned advisers that they were not expected to express their own views.

Advisers who adopted the adversarial mode seemed to have varying objectives. Some visiting parties contained one or more members with an unequivocally aggressive style; one adviser was seen in that light by the unit, but not by the observers. Some advisers holding strong views about how research ought to be conducted were seriously critical of a unit's work. They attacked weaknesses in specific projects that represented more general problems. The assertive style of one represented an attempt to make a fundamental challenge, and to make a significant impact on the value system of the unit concerned. Others seemed more simply to believe in the value of criticism as a mode of intellectual test - they were there to test the limits of the competence of the researcher and of the strength of the research. In one case, the same adviser grilled intensively a researcher of strongly established reputation, but was mild and unthreatening to an inexperienced worker with an uncertain research base: the weaknesses were already exposed.

Presentation reflected the units' range of research and extended into border areas between, for example, research and dissemination and research and development. Questions aimed at determining whether basic scientific criteria were being met. Was the logic of the inquiry sound? Were the aims clear, and the methodology appropriate? What evidence was there that the researchers had the capacity, time and resources to complete the research successfully? Were the confidence limits of the research clear? Were the data sufficiently controlled for the results to be generalisable?

Such criteria were evident in questioning of research judged to be of widely differing quality. In one of the units complex work on the natural history of psychiatric illness, entailing the relationships between personality and social indicators, and between subjective and objective evidence, was being attempted. It was also studying the diagnostic processes in general practice and outcomes in social work, where the nature of evidence, the criterion of validity and feasible degrees of precision and certainty are highly complicated questions. In other words, in areas susceptible to a wide range of 'hard' and 'soft' approaches, only 'hard' criteria were applied.

In two visits the theoretical bases of work presented were examined. In all cases they were judged as requiring some strengthening; and, in a few, radically reappraised. In one unit, the most consistently strong in its methodology, relatively little attention was given to basic scientific criteria. The main challenges were on theory, scientific originality, initiative and independence, and the scientific importance of its work. Pressure was applied where possible weaknesses were perceived.

Sometimes complex issues were raised, either by the unit in response to questioning from the visiting party, or by advisers. For example, one project originally intended to move to curriculum building on the basis of role analysis of a profession. This revealed profound differences about the role analysed and, in consequence, doubts about the mandate for researchers in curriculum building.

Again, two units working in action research were committed to strengthening research's potential to effect change. Advisers' doubts about this method reflected the view that science ought to legitimate rather than to stimulate change. But these differences of value were not exposed. Elsewhere, there was dissent between the unit and their visitors about the legitimacy of their kind of research, about the nature of the database, and about the reliability of its findings. These basic differences remained unexplored on the day of the visit.

Thus encounters between researchers and advisers were largely concerned to establish whether basic scientific criteria were met. The visitors, however, raised their expectations as they became convinced of basic competence. There is some indication that time-limited and adversarial evaluation closed off important material. More exploration of complexities and value differences might or might not affect judgements; it could, however, make them more acceptable and usable. A senior researcher spoke of a deep sense of frustration that he had not had the opportunity to clarify what he saw as a crucial divide between himself and the visiting party on the nature of his field and the basis on which work should go forward (IR (80)/30.

The project presentations were followed by private meetings of the visiting party. These had two purposes: the assessment of the scientific merit of each piece of work and making judgements and recommendations on the scientific problems, functions or management of the units as a whole.

Almost all visiting teams balanced criticism with support. In some, the visitors considered alternative lines of attack on research problems and means of strengthening the work in progress. Comments about the functioning of the unit as a whole were future oriented. Priority might be given to identifying general problems and to thinking about how they could be overcome. They considered whether the unit's time was appropriately allocated between activities, whether the size, experience, deployment, differential influence and conditions of employment of staff were appropriate, and whether the range of work was conducive to quality. Then they made recommendations accordingly.

Decisions were influenced by the DHSS's commitment to the unit, by its historical scientific strength and institutional base, by some recognition of the need to pursue disciplined approaches to implementation and dissemination of theory, even if many did not meet the criterion of scientific rigour, and by attention to the stage of development reached. The problems of evaluating and changing practitioner behaviour in general practice, nursing, social work and in the education and care of the mentally handicapped were aired. One visiting team acknowledged the problems of beginning work in a virtually unmapped area. However, while these problems were recognised, in only one case were the implications fully acknowledged. The limitation of random allocation studies in circumstances where the internal process of practitioner-patient systems remains ill-analysed, the intellectual and management problems of researchers in a new field, and the minefield of the theory-dissemination-practice relationship - some of these factors were not in the end allowed to weigh heavily against the requirements of scientific control. It was not that the evaluation process was too rigorous but that it did not allow for appreciation of complexity and therefore risked narrowing the research base.

One visiting party stood out for the range of the scientific issues raised. It discussed conditions for encouraging researchers to grip complexities. It considered the implications of movement into a post-paradigmatic phase, of challenges to established paradigms, and of the problems of building up new ones. It concerned itself with the compatibility of scientific excellence and external commissioning. It discussed the relationship between epidemiology and social science. Both scientific and social criteria weighed heavily with it.

The dynamics of different visiting parties varied. A chairman might structure the discussion meticulously, inviting advisers to give judgements on the presented work in strict order, and himself giving a succinct summary of each section of the discussion. Alternatively, a chairman might confine himself to identifying or simplifying issues, and make no attempt to summarise the views expressed, except briefly at the end. The most orderly discussion evinced little controversy between advisers, although conflicts of perspective were evidently present, and emerged clearly when the report of the day's events was tested by challenge from the unit. However, in a less structured meeting it seemed that some advisers were allowed more say than others, and one occasion was dominated by one adviser. In only two meetings did conflicts of view on projects strongly assert themselves.

Shared values seemed more important to advisers than conflicting ones. Interviews with advisers following visits for the most part confirmed this impression although there were dissenting voices. One adviser saw a danger that Chief Scientist's visits could collude to reinforce illusions of certainty; another believed that while so many uncertainties remained in DHSS policy on research units, Chief Scientist's visits were weakly based and given undue importance. Many were concerned that the complexities of evaluation were not fully taken account of in the visit procedure.

Except in one case, formal feedback of the visitors' conclusions on the day was negligible or non-existent. After a few visits the OCS ruled that advisers should avoid giving the director any intimation of the conclusion of the visiting team, whether favourable or unfavourable. Only four researchers expressed satisfaction with feedback on the day, while advisers were more evenly divided. Researchers' main complaint was that lack of feedback created a sense of anti-climax after the effort they had invested.

Most advisers thought the judgement too important to be conveyed orally before there had been time for reflection. Others thought that a judgement should be given partly on the grounds of openness, and partly to concentrate the minds of advisers and to ensure that they accepted the consequences of their assessment.

The report of the Chief Scientist's visit

The report of the Chief Scientist's visit to a unit was compiled by the DHSS liaison officer. There was an established format. Participants in the visit were listed, and the report then followed the main events of the day. The purpose of the report was, in the words of a research manager, 'to inform the Department, not to develop the unit.' (IM (80)/1). Submission of the draft report for factual comment was seen by the Department as a simple clearing operation: the director was asked to make substantial comment only on the final version. This was regarded as a report to him or her and he or she was to decide whether it was read and responded to by the whole unit. The directors who raised the strongest objections to the report released the first version to the unit as a whole, and sent substantial comments to the Department without waiting for the final version.

The rationale for the departmental procedure is understandable. This document was an extremely important one for both the unit and the Department. The need to produce a version agreed on fact and not open to misinterpretation was clear, as was the need to separate the document from comments upon it by the unit. However, the Department perhaps underestimated units' investment in the Chief Scientist's visit and their anxiety during the waiting period for the verdict: 'waiting for a biopsy report' as one adviser called it (IE (80)/3).

Also units might consider judgements to have been reached upon inadequate evidence, or to have exceeded the terms of reference of the visit. If they were damaging the directors did all they could to have them deleted from the report itself. That was the authoritative document. The appended document containing the unit director's own comments had no comparable status. In the absence of an appeals procedure, however, directors could only lodge objections and the Department's carefully worked out procedures became prey to conversion from an adjudication to a bargaining process.

The procedures were severely tested. While two units were broadly satisfied with their reports, and one was critical of a number of aspects of theirs, the other two had serious objections. Some judgements were challenged on the grounds of inadequate evidence and of misunderstanding of the nature of the research.

When one unit raised serious objections, advisers reconsidered the report. Some stood by the original version but comments from others reflected significant reservations. These were about the scope of and procedures for Chief Scientist's visits, their examination in many cases of only part of a unit's work, and the exclusion from them of policy-makers. Others were critical of the principle on which reports were compiled. And while no members of the visiting party conceded errors of judgement one thought methodological problems had not been fully acknowledged, and several expressed concern about the balance of judgement emerging from the day's proceedings. However, the report was substantially changed in only one respect. The consensus reached on the day having been shaken, there was then no clear procedure for the Department to follow.

The objections of a second unit were dealt with slightly differently. They were considered by the liaison officer and the Chief Scientist and submitted only to the visiting chairman. Objections that judgements had exceeded the visit's brief and that other judgements were based on inadequate appraisal of the material presented were rejected by the chairman who was supported by the Chief Scientist. In consequence the report was revised by the liaison officer to incorporate the unit's factual and interpretative corrections only. It was resubmitted to, and cleared by, the rest of the visiting party at this point, and then referred back to the unit director who made fundamental criticisms of the revised principles and procedures of Chief Scientist's visits, noting that the visiting party had not in practice adhered to them. He also dissented from many of the judgements made. On this basis the customer review of the unit began twelve months after the Chief Scientist's visit.

These events indicate the serious difficulties that could arise. If the Department conceded to the objections to their reports raised by units, particularly more influential ones, it might be open to charges of yielding to power; if it did not, then a sense of injustice would persist.

A particular problem raised about the report stage concerns the basis on which reports were compiled. Three of the advisers who qualified their views to meet objections by the first unit distinguished between a record of the day's proceedings and a report summarising judgements of the quality of a unit. Four research staff from another unit thought that an account of the day's proceedings was an unsatisfactory vehicle for recording serious judgement. Representatives from three units objected that the judgements recorded went beyond the data available.

Representatives from four units were highly critical of the quality of their reports. Some researchers had expected reports to be for the benefit of the units as well as of the Department. Even the most detailed did not come near to satisfying the desire of the researchers to learn substantially from the reports.

The concept of scientific merit and the process of evaluation

A central issue that emerged concerned the epistemological assumptions inherent in the process of Chief Scientist's visits. The institution of the visits reflects the view that collaboration between scientists and policy-makers could also be contaminative. Under the procedures adopted in 1979 these visits were concerned solely with the assessment of scientific merit. Although scientific advisers applied a strong common core of criteria, the concept of scientific merit was by no means the same in all visits. Generally, the stronger the researchers and the more powerful the unit, the broader were the criteria applied.

Some advisers interviewed for our study elaborated the complexities of judging the quality of work, if that judgement were to extend beyond basic technical competence. Several stressed the personal nature of scientific judgements; one spoke of the need to combine individual values with a deep knowledge of the stage of development and intellectual strength of one's field, and of how the collective view of scientists in particular fields is continually shifting. The differential development of concepts and instruments of measurement between fields was emphasised by three of those interviewed. In a pre-paradigmatic field, unwarranted assumptions were sometimes made about what was taken for granted, about appropriate methodology and about criteria of quality. Lines of controversy might not be clearly demarcated. The balance to be struck between theory generation and theory verification could be more than usually uncertain. Tight control of data in evaluating, for example, care of the mentally handicapped might be useless if the data were inadequate to develop the concepts of quality of care in this field.

Anxiety was expressed by some advisers as well as by the directors that the selection of advisers did not take account of the criterion of scientific coherence. Others considered that the assessment process was subject to excessive reduction of evidence and that assessment of the unit as a whole was unsystematic and even arbitrary.

The reduction process was clear and particularly pronounced in the case of multiple-funded units. We have already noted the restricted scope of work presented on the day and that it was oriented towards the examination of discrete areas. In the words of one adviser, the assessment was 'modular' (IE (81)/2). The division of labour for advisers focused their attention on what they particularly examined and the report summarised their comment, which was likely to derive from that focus.

Some of the sense of injustice among unit directors derived from their belief that judgements were made in areas where the visiting party had inadequate evidence or where there had been no opportunity for full exploration. One director thought that the strength of the unit's academic networks and influence was understated, two that the links between research in the unit had been seriously underestimated. Another considered that the constraints upon its scientific base which derived from its overall function had not been appreciated. And one thought that the balance of the unit's work had been misunderstood.

The decision to make 'scientific merit' the sole focus of Chief Scientist's visits threatened the basic assumptions of some research units. Such units, often with the encouragement of the DHSS, concentrated upon policy-relevant research that made no pretensions to influence the course of science. They feared that their work would be subjected to criteria quite different from those prevailing when they began it.

Some of the presented work certainly tested the definitions of science: for example, action research in the development of nurse training, and the development, trial and evaluation of educational packages for the mentally handicapped. The research units themselves differed in their thinking about scientific rigour and policy research. Some were clear that research must remain within the established parameters of controlled investigation. Their task was to develop instruments of measurement that could extend their work without transgressing those parameters. They emphasised the unity and the boundaries of scientific method (Nagel, 1961). Others had been more influenced by the critics of positivism and their impact on definitions of social science and social research. Their work reflected too a broad-based anxiety about the efficacy of research undertaken to evaluate policy-relevant programmes of all kinds.

Some researchers in the units whose visits we observed were vulnerable to criticism that they were falling short of scientific standards. Some units worked within terms of reference that were neither science nor research: for example, an epidemiological reporting system, the provision of a referral system for mentally handicapped babies, and the building-up of a research information service. Those whose research, service, dissemination and information or communication functions were integrally linked with each other found that the scientific visiting parties mostly rejected the essential nature of these links except where they were incorporated within research projects.

It seems then that while, with very few exceptions, the researchers regarded themselves primarily as members of the scientific community, they had different views about science and about the relationship between science and policy research.

Scientific merit and policy relevance

Most unit staff considered that their assessment had been confined to science, but a significant minority thought that criteria other than those of science were applied. Testing the strength of the principle of separating scientific merit and policy relevance entailed two questions. Ought science to be assessed by scientists separately from the policy-makers whose concern is with its usefulness to them? Can science be assessed independently of an understanding of the policy questions that it is tackling and of the constraints upon it? The judgements of some of the participants in our study were clear. One spoke of the danger that 'alpha judgements of scientific advisers could be subjected to the beta assessments' of policy-makers if they helped to determine scientific merit (IE (80)/2). He and others considered policy and science to belong to different universes, although views about their respective nature and contributions varied. A researcher in mental health saw science as generating ideas about ends, while policy supplied the means. Conversely, a statistician working in the field of professional education considered that the research questions derived from policy; science must see that the questions were properly tackled.

These responses highlight important differences within health and personal social services research. Three distinct categories of research can be described in which the relationship between science and policy is different. There is, first, research upon, for example, the outcome of drug prescribing or upon the epidemiology of diseases of the central nervous system. Such research has important implications for policy-makers but the questions, the causal theories, the hypotheses about treatment, and the methodologies all belong to established fields of medical and health science: a separate system from that of policy.

Such work can be contrasted with the application of established disciplines, theories or methods to a policy problem; for example, the application of micro-economics or epidemiology to resource allocation or assessing the impact of social policies on the health of children. Here science can be seen as different from policy but as instrumental to policy needs. It is different again from a field such as nursing or social work or family policy research which is derived in part from policy questions. (What benefits or services operating under what manpower policies might promote family cohesion? How should they be allocated?) Research in such fields is usually multidisciplinary. It concerns the relationship between systems and individual behaviour or conditions, and there is no established discipline or research field with a prior claim on it. It is domain-based research (Trist, 1972) and not separate from policy. Good research in such work depends on the development of both method and theory and their interdependence.

Answers to such questions as 'Can the disabled be integrated into society?' entail concepts of integration and the values with which they are connected, as well as the constraints upon their implementation. This requires access to the thinking of, amongst others, policy-makers and professionals. Empirical data needed for explanation are to be found in the settings within which the policy-makers work. Policy-makers also conceptualise problems in such a way as to test and present alternatives to those identified by researchers. It is true that they may reduce, or take out of context, the products of scholarship. They may also offer legitimate challenges and elaborations. Two policy-makers described their close collaboration with a research team working on the preferences of disabled people. The DHSS had persuaded the researchers to focus part of their work on the 'cash against care' issue. The policy-makers considered that their questions and understanding of the policy issues had contributed to formulating objectives and their alignment with methods in the project (ID (80)/19). A later encounter between the researchers and members of the DHSS showed how administrators, professionals and economists could combine with researchers to evaluate a project at the pilot stage. The boundaries between 'government' and 'research' emerged as relatively open and shifting.

Most of the advisers interviewed were unhappy about the exclusion of policy issues from the Chief Scientist's visits. Some thought a division between scientific content and the relevance of research to be artificial. Policy researchers would incorporate criteria other than scientific merit which is not always a sufficient yardstick of the quality of policy relevant research.

The advisers in all visits emphasised the criterion of 'importance' for scientific or for policy development. Two advisers noted that motivation to influence social development affected the degree of importance to be placed on researchers' work. Scientific quality was thus promoted by social motivation and values other than the 'purely' scientific might be entailed.

Some advisers who accepted the separation of scientific merit from policy relevance in principle nevertheless thought it unworkable. One considered that the degree of certainty required, the available money and the relevance or urgency of the research, had always to be balanced. Science was not an independent variable. Others, too, argued for a greater emphasis in the visits upon the context of research, even if policy and science were seen as separate concerns.

It has been contended here that different forms of policy-relevant research entail different relationships between policy and science. In some, criteria of policy relevance and scientific importance are difficult to keep separate. In most, judgements need to be informed by an understanding of the economic, policy and scientific context. In some cases the knowledge systems of advisers will not be distinct from those of policy-makers; accordingly the judgements made by advisers may extend to the formulation of policy problems as well as to the methods by which they are tackled.

The principle that the assessment of scientific merit and policy relevance ought to be separated is not destroyed by these arguments. However, an over-rigid application of it may lead to distorted judgements.

The concept of scientific merit and the adversarial mode

> The practice of scientific method is the persistent critique of arguments, in the light of tried canons for judging the reliability of the procedures by which evidential data are obtained, and for assessing the probative force of the evidence on which conclusions are based. (Nagel, op. cit)

The adversarial mode adopted in visits is entrenched in the scientific tradition. The development of science is often said to be dependent upon scientists' continuing exposure to criticism and argument by their peers. Science is, it is argued, essentially public knowledge that is legitimated in so far as it can be shared through convincing empirical evidence and argument (Ziman, 1968).

The automatic self-regulation of science through the operation of universally accepted norms has, however, been convincingly challenged (Mulkay, 1979). Yet both the MRC's and the Chief Scientist's visits are broadly based upon this idea. Participants in our study had differing views about the mode of evaluation and the scientific norms entailed.

Arguments in favour of the adversarial mode were that a basic requirement for the promotion of good science can be met, namely the elimination of unwarranted error, resulting from weak logic, uncontrolled data producing uncertain confidence limits, or ill-defined terms and inadequate analysis. Beyond this, challenge can, by testing theoretical assumptions, methods and findings, provoke new lines of thought. Judicious selection of questions can enable broad as well as specific judgements of quality to be made. Strict time limits (one-day visits) produce both concentration and economy, and economy is important if the best minds are to be recruited for the task.

One criticism of the adversarial mode is that its efficacy depends on the stage reached in the field, which affects the degree of consensus about criteria of evaluation. In the absence of consensus, judgements using particular criteria can appear to be imposed through power rather than through resolution of differences, for which there is no time in this mode of assessment. Alternatively, expectations of tough testing on traditional scientific criteria may discourage researchers from exposing the complexities of their work so that they undersell its importance and fail to test the criteria involved. Again, some argue that some research requires more time to explore the subtleties that provide its essential value.

What support for these arguments, which are distilled from interviews in our study, was provided by observation of the Chief Scientist's visits? The visits provided a forum in which logical weaknesses and methodological problems could be exposed. Visiting parties recommended that work with a base judged to be unsound should not be further supported, and that exhausted lines of research should be brought to an end, although criticism was, in some visits, more consistently sustained than in others.

Some visits, then, fulfilled the function of eliminating error and of redirecting work of limited potential, making proposals on better use of resources. The impact of such challenges upon the units could not as yet be clear to us. Advisers used the work presented to them to make more general judgements about units. Several of the arguments for the adversarial mode of assessment were, at least in part, supported in the events of these five visits. Our experience was generally that the strongest work was most strongly tested. Neither was there wholesale rejection of relevant work that was not sufficiently strong if there were other arguments in its favour.

Criticisms of the evaluative style of the visits focused on the illusion that any such occasions could ever be wholly rational, objective or directed towards a single purpose. It was said that what is tested in such encounters is not necessarily scientific ability but adversarial skills or the capacity to think under stress. The complexities of group judgements of scientific quality were thought to be insufficiently acknowledged.

The behaviour of researchers could impinge upon the process and because of the mass of material and issues confronting advisers, a particular piece of work could attract disproportionate attention. The processes through which such selection were made might not be scientific. Nor, it was thought, did Chief Scientist's visits fully conform to notions of peer review. Some researchers in our study experienced their visit as a review by equals; most did not. Scientists do not feel they belong to a republic of equal citizens. Visiting parties were constituted largely of professorial or very senior researchers, while the seniority and experience of unit staff were widely varied. The multiple focus of most units, and the application of general scientific as well as specialist criteria, ensured that reviews were not peer in the second sense: review by those working in one's own specialism. Finally, the visiting parties combined scientific authority with the power of a system whose decisions could mean life or death to the units. Those who constituted a peer group for unit members were also part of a governmental process. Advisers from new research fields may have been anxious to assert that they were peers of their fellow advisers. Their membership of a visiting party might symbolise recognition for themselves and the work they represent, and perhaps help them in the struggles in their field about dominant norms and identities. Advisers rejected the view that such motives propelled members of their visiting party 'to perform' before their colleagues, but acknowledged that it could have been so.

Many participants in our study thought the adversarial mode basically sound, and proposed changes that would strengthen, not abandon it. Others proposed some shift towards an interactive style of evaluation so that advisers would be more exposed to the researchers and less to their colleagues on the visiting party. It might make it more difficult for the visiting party to assert a consistent set of scientific norms across the unit. Instead there would be stronger emphasis on reaching understanding of units' own purposes and perspectives, and less stress upon their conforming to a more generally defined, and perhaps too restrictive, set of criteria.

Modelling alternatives

Our discussion has indicated alternative models for Chief Scientist's visits. The possible evaluative *structures* that emerged include the *discrete/modular* version in which the individual project is the focus of assessment; the *holistic,* in which the unit is assessed primarily as a whole; and the *contextual,* in which the unit is assessed within either its scientific or its policy context or both. Evaluative modes range from *authoritarian/adversarial,* in which the full power and authority of government and the scientific community support the critical examination of units' work; through *egalitarian/adversarial* in which the contest is made more equal; to *interactive* where the substantive power and authority are tempered by the value of mutual exploration. These evaluative modes and structures may be concerned with *scientific quality* or *technical competence* in work categorised as *policy-relevant science, policy-instrumental science or domain-based policy research.* Such work may be centred in the traditions of scientific method or at the boundary between science and other forms of knowledge.

The epistemological issues in evaluation cannot be wholly divorced from questions of authority and power. The predominant evaluative mode of Chief Scientist's visits worked well in many respects. But it is likely to be most acceptable when there is a consensus about criteria or equality of power and authority between assessors and assessed. Policy-relevant research is too diverse to be subsumed under one method of inquiry, and there cannot be one authoritative view. These arguments may be rejected by those whose prime concern is with scientific standards, the authority of government and accountability. However, they can be further considered as we analyse the second stage of the review of units: that by the customers.

Chapter 10: Review of Units and Policy Relevance: the Customer Review

Introduction

OUR DISCUSSION of the customer role in the review of research units will echo some of the themes that have already emerged. Once again the monolithic structure of the Department emerges as an illusion. We discern different patterns of decision-making in which the relevance of a unit's work had different meanings. Separation between the roles of scientific advisers and policy-makers was emphasised in a system directed primarily towards collaboration. We take up the epistemological rationale for differentiation and the extent to which assessments by customers and scientific advisers were in the end collated. But we reflect too on how the revised procedures for review asserted the Chief Scientist's executive role in the DHSS.

The assessment of policy relevance

The guidelines drawn up by OCS (IM (79)/23) for the review of units' work following the Chief Scientist's visits described an apparently simple, separate and internal exercise. Once the scientific standards of a unit were judged to be satisfactory, its 'past and prospective' relevance to customers was reviewed so that its future could be determined.

Decisions were made by the policy-makers and OCS: RLG chairmen or the head of the appropriate customer divisions where no RLG existed, their professional colleagues, the Chief Scientist, unit liaison officers and the senior OCS administrator. Recommendations were put to the RLG where there was one. If they were confirmed, the Chief Scientist would then authorise a revised or 'rolled forward' contract with the unit, and a detailed programme would be worked out. The OCS's role was strengthened by the separation of the review into two distinct parts. Not only did the Chief Scientist have executive responsibility for the commissioning of research programmes, but the OCS was the only group involved in both parts of the exercise. It thus had a key coordinative role and it did not have to start the evaluation from cold. The RLGs, however, were allocated a secondary role. Research units had no direct representation in the proceedings.

Expertise and accountability were asserted in the procedures. Relevance was the expertise of policy-makers and science that of advisers and researchers. And accountability for research commissioning rested firmly in the Department. Its exercise was not to be influenced by people external to the Department who might assert criteria other than relevance. The RLG's secondary role also avoided double scientific review of a unit and maintained the authority and independence of the judgement of the visiting panel. But the equivocal attitudes to the RLG demonstrated that the principle of collaboration between scientists and policy-makers did not fit comfortably with those now asserted.

Policy-makers might interpret the task of determining relevance variously. They might simply consider whether the unit's work had been useful to them in the past, and whether researchers' plans accorded with their needs sufficiently to extend the contract. If not, they might want to put different suggestions to them. Alternatively, they might undertake a much more strategic exercise in determining the role of the research units. In the case of a large unit that might entail a major reappraisal of customers' research policy. In each case the review procedures denied a full strategic role for the RLG and lost the opportunity to develop collaboration between scientists and policy-makers in an important way.

A mixture of interpretation, principles and role allocations asserted themselves in the review process. For some policy-makers a customer review of units had little meaning, and they assumed that their role would be limited. Some were not sure whether the report of the Chief Scientist's visit would be made available to them, while others had not expected that they would need to read it. Others were more actively involved and treated the Chief Scientist's report as essential to the assessment of the unit's relevance to their work. However, even amongst this group, not all had realised that there were revised procedures for unit reviews.

Whatever the patterns of decision-making adopted, the processes were lengthy and complex in our cases. Fourteen months after the first of the Chief Scientist's visits to the units, the contract of only one unit had been rolled forward. In no case was the process of review complete, and in two the customer review had not yet begun, although the reports of the visit to all the units except one were cleared within five months of the visit.

The different roles allocated to the RLGs, together with the different styles and emphasis of the communications between the unit liaison officers and their RLG chairmen, added up to some quite different patterns of decision-making. In one case, communication was informal and the action taken implied a collective approach between the liaison officer, the customer divisions and the RLG. Discussion in the RLG steered clear of making a second review of the unit's scientific standards, but the implications of the report's comments and recommendations were considered in detail. The report was thus fully incorporated into the process.

In another case, the roles and tasks of the liaison officer, the client group and the RLG were clearly differentiated. The liaison officer worked with the unit, and expected that the client group would produce its own recommendations once she had provided a clear contextual framework. The client group involved the RLG but ensured that the RLG discussion focused on the unit's own 'forward look' and not on the report of the Chief Scientist's visit itself.

In the third case, roles were again clearly differentiated, but the liaison officer took a more active part in structuring the decision-making process and in working out the short- and long-term implications of the Chief Scientist's report. In the event, the opportunity to reflect upon the unit's future contribution to research strategy in the RLG was made but not fully exploited by either the policy-makers or the scientific advisers.

The first unit was a post-Rothschild creation of a close-knit professional division and the liaison officer was a professional. The unit was small, all its research done for the DHSS, and it was structured round projects. The customer division and the RLG had been closely involved in its progress from the beginning. It was assumed that it would constitute the centre for that field of research, and its creation was seen as an important element in strategies to develop the profession.

The other two units established themselves as centres of research long before Rothschild and, although neither was accorded the status of a DHSS unit until after the new research management system was set up, they had already had several years of working under contracts with the DHSS. They were also multiple-funded units. Their work had thus developed separately from the RLG strategies.

Their liaison officers were professional research managers, but without professional links with the policy divisions concerned with their units. One had a background in social research and the other in general practice. Their units were in mental handicap and mental health. It was thus natural that the roles of policy-makers and liaison officers would be more clearly differentiated in such circumstances.

But there were some similarities in the approaches to the review of these three units. Roles allocated to the RLGs although varied were in all cases significant. Policy-makers' commitment to RLGs was confirmed in this study. In all three cases the unit review was perceived as an opportunity to reappraise the role of a unit in the research policy for the division. The RLGs were involved to a lesser or greater extent in a strategic exercise in which the collation of the views of scientific advisers and policy-makers was an important factor. The intentions were not fully realised. It was hard for policy-makers to give priority to long-term work of this kind. The existence of RLGs together with the need to scrutinise the work of researchers more closely in a period of economic constraint provided a useful impetus. But the creation and sustainment of productive dialogue with scientific advisers and researchers require more than this.

Outside the RLG areas, there were likely to be few policy divisions with explicit research strategies against which a unit's past work and future interests might be set. Therefore it might be difficult for either the OCS or the unit itself to collate research interests with policy needs in any consistent or developmental sense. Such difficulties were exacerbated when a unit's field of work crossed policy division boundaries. There was no established coordinating machinery through which a coherent research programme geared to departmental needs could be worked out in these circumstances. Thus in the absence of RLGs, the customer review process was likely to be more idiosyncratic and limited with the onus of collating the views of different customers resting with individual liaison officers.

Did customers reach different judgements from scientific advisers?

From the process of the customer review we now turn to substance and our evidence of differences between the judgements of policy-makers and scientific advisers.*

Our earlier discussions of the nature of policy research suggested that the criteria applied by policy-makers and scientists were not wholly separable. A policy-maker was surprised that the report on a unit was not more 'scientific' (ID) (80)/21). Policy-

---

* This account is based on interviews with customers of research units, observation at departmental meetings and scrutiny of departmental papers. Details of the numbers of interviews conducted are noted in the Appendix to Chapter 8 of this book.

makers, however, identified two specific areas in which they differentiated their approach from those of scientists. Two made a distinction not explicit in Chief Scientist's visits: between research that is technically reliable and research that is sophisticated or has scientific merit. Their concern was primarily with the former, but they assumed that the interest of the Chief Scientist was in the latter. Other policy-makers discussed the problem of implementing research, in particular where it would require excessive resources. The example they gave was one where the scientists' problems might be generalisability. However, they feared that the research might be esoteric and large-scale implementation impracticable. But such different criteria might produce apparently similar judgements between the two groups.

Emphasis on the soundness and relevance of research was reflected in customer views of the function of units. Most customers took an instrumental view. One believed that units should constitute centres of excellence and another emphasised their diverse roles. The more general view was that units should answer the policy-makers' questions, that their work should be structured round individually commissioned projects, and that it should be practical.

However, other policy-makers gave thought to the degrees of freedom allowed to researchers. One policy-maker was concerned that her division had managed to commission very little research of direct use and that contracts with units should be tight and clear. But as a person who was interested in and used research, she also had liberal views about it, doubting whether government should be involved in its management at all. Another group wondered whether it had exercised too tight a control over its units and had thereby stifled initiative. A third, who had worked closely with one of the units, believed that expecting a unit to meet the expressed needs of the Department need not encroach upon researchers' freedom, nor prevent units developing as centres of expertise, if not excellence. Linked projects and extended work could be extremely valuable to the Department. Structured collaboration could set up free and productive dialogue between policy-makers and researchers. But in such circumstances, independent scientific appraisal could be a valuable safeguard against complacency or excessive 'cosiness'.

Our study yields a complex picture of the relationships between the judgements of policy-makers and scientific advisers. The review of one unit was particularly overshadowed by the issue of relevance. The Chief Scientist himself, the unit liaison officer in the briefing for the visit, and the RLG member on the visit, were all concerned with whether the work presented by the unit was within the DHSS contract as well as with its scientific merit. In other respects, the judgements of the scientific advisers and the policy-makers were compatible but distinct. The scientific advisers were critical of the science of some of the research, the value of which was doubted by the policy-makers because it assumed a policy framework which was unlikely to be adopted. The scientists' judgement was reached at a time when the policy-makers were beginning to clarify what they might like this unit to tackle. But there did not seem to have been any connection between these two occurrences. Nor was there a discernible attempt to orchestrate the different orders of criticism into a concerted condemnation of the unit. Rather, efforts were directed towards establishing a base from which a more mutually beneficial contract could be made. Elsewhere, policy-makers more actively welcomed an independent scientific appraisal of the unit's work as a means of testing the validity of their own views. One unit was newly launched by the policy-makers themselves and they were reassured to have its value satisfactorily supported by scientific judgement. Their judgements of some of the individual projects of the unit were not very clearly distinguishable from those of the scientists. They also believed that the unit had received support from the visiting panel for its involvement in the politics of some of its research. It was significant that the support emanated from the scientific adviser most convinced that the distinction between scientific merit and policy relevance was false.

In a second case, the customers valued the unit concerned, but felt that it had acquired an unassailable but not altogether legitimate authority. They considered that the Chief Scientist's report had made possible a more critical approach than they had previously adopted. Some of their criticisms were similar to the scientific judgements; others were in direct conflict. Work that they considered esoteric was generally held by scientists to be of high quality. Their views about the unit were also imbued by views about the importance of maintaining boundaries between research and policy-making. They felt that in this unit scientific and political authority had sometimes become confused.

So far, then, we have built up a picture where the divide between scientists and policy-makers was not enormous. The distinction between assessing scientific merit and policy relevance had not been wholly sustained in the review process.

But judgements could not in all cases be brought into alignment. In two cases the scientists were critical of units considered by the policy-makers to be important and valuable to them although they stopped short of making recommendations in direct conflict with the Department. The first unit had a high reputation. It had worked extensively on methodology in the field of epidemiological research, and it had developed close working relationships with customers within the Department. The Chief Scientist's visiting party did not dispute its reputation, but identified limitations. Their judgement was rooted in a broader concept of scientific quality than technical reliability and in effect they said that the Department needed work that was more than methodologicallv sound. In questioning the compatibility of scientific merit with responsiveness to problems formulated by policy-makers, they were also applying scientific criteria that did not easily accord with an instrumental view of policy research.

When the review by the policy divisions began, an instrumental view of research emerged, and along with it a view of the political function of research. Objections were raised to discontinuing an area of research on the grounds that its results were strengthening the Department's arm against international pressures to take a particular line of action. In another case, withdrawal of support from research in a particular part of the country was, it was argued, politically unwise. Moreover, professional and scientific representatives from one division felt strong enough to challenge the judgements of the Chief Scientist's visiting party on a particular line of research. And although the visiting party's recommendations for a reduction in the size of the unit, made in the name of scientific coherence, coincided with the Department's need for financial cuts, the Chief Scientist's arguments carried relatively little weight against those of policy-makers based on policy needs and political issues. In yet another field a policy division without an RLG was developing its own ideas about future research bases in direct contradiction to a recommendation of the Chief Scientist's visiting party.

In the second case, the unit provided a data base heavily used by two policy divisions, the continuation of which was to form part of a new study which these customers considered essential. Scientific criticism extended to the adequacy of that data base. Here policy-makers, while not disputing the scientists' judgements, did not fully accept their implications. In general statements, customers took the position that if the Chief Scientist's advice was that the science of a unit was poor, they would not continue to support it. But if the advice was critical but not devastating, and if the work was important to them, they would contest the judgement and their views would carry weight.

The studies that were the subject of this dispute went through a preparatory process that might have been expected to avoid the problem. A steering group was set up in which policy-makers, statisticians and the OCS were represented. A compromise was reached between the needs of the policy-makers and scientific viability, and a proposal went forward for commissioning. However, it was then subjected through the OCS to the normal commissioning procedures which entailed independent scientific scrutiny. Two systems of appraisal were thus brought to bear upon it, with the clear possibility of conflict between them. The outcome was that the unit was commissioned by the Department to undertake the project, but against the 'better judgement' of the Chief Scientist. Policy-makers felt that they must insist they considered imperfect data preferable to none.

Such decision-making processes were a clear example of the Chief Scientist attempting to impose, in his view, the more valid principle of the separate assessment of scientific merit from policy relevance in place of compromises between scientific rigour and policy needs. This attempt was the more explicable by reference to the position of the Chief Scientist in the Department.

The Chief Scientist managed the research programme and authorised the making and renewal of contracts. But his formal authority *vis a vis* the policy divisions was limited. It was assumed that disagreement would be settled by negotiation within the Department.

In such circumstances the revised procedures for review can be seen as part of a wider policy to assert greater scientific authority within the Department. The authority of science is less than formidable when set in direct opposition to the power of a policy-making institution which makes judgements upon multiple criteria.

Authority, influence and the revised review procedures

The research management system set up in response to the Rothschild recommendations emphasised collaboration between scientists and policy-makers. By 1978 policies emerged which entailed assumptions more like those of Rothschild's customer-contractor principle. These were that policy-makers and scientists belong to quite different knowledge systems and institutions and carry their respective roles best when clear divisions of expertise and tasks are maintained; that policy-makers are well able to formulate their research needs, and that there is a strong, independent scientific community with the research capacity and experience to meet those needs under contracts with the Department. A predominantly internalist view of science was asserted in the principle that science and relevance must be separately determined. The authority and influence of science were perceived as dependent upon the adherence of those working under its aegis to the standards set and sustained through its own systems, and pursuit of this principle would enable the Chief Scientist to establish the strength of his position in the Department. Further, just as scientific judgements ought not to be contaminated by the views of policy-makers, so policy decisions should not be infiltrated by scientists. Policy-makers too were to assert their proper role.

Thus clear allocation of authority and responsibility within the Department was expressed. The organisation of the review then demonstrated that responsibility for asserting the standards required of its research units was held firmly in the Department. The Department had established the principles and the organisation for review and retained power over it at key points: in particular in the selection of the Chief Scientist's visiting party, in structuring the choice of work for presentation, in the management of the report of the Chief Scientist's visit and by the internal nature of customer review. It was assumed that the centralised authority structure of the review was underpinned by a consensus about objectives and standards, and that procedures were best mediated through relationships and a common language rather than through rules or rights. Participants in the Chief Scientist's visits and the customer review clearly attempted to implement the principle of a separate assessment of scientific merit and policy relevance. In the case of both, absolute separation of the two elements was found in practice to be difficult to sustain. As the procedures were worked out, questions of power and authority within the Department appeared not wholly soluble by division of task and experience. Yet the revised review procedures represented a considered attempt to assert the separate authority of science and of government.

An alternative perspective

Are there other options for the process of review? The DHSS research units in health and personal social services were from their inception subject to multiple and changing expectations. They were engaged in research areas shaped by competing paradigms. Their work reflected problems in the relationship between the natural and social sciences and the conflict between positivist, interpretative and critical stances in social science.

The nature of the research raised questions about science as the sole yardstick against which it should be measured. Some work was shaped by and rooted in science, and this must continue to be reflected in evaluation; but some was rooted in the policy problems it sought to understand or resolve. Much of it could not realistically be assessed without reference to the policy field concerned. Some of these fields, such as nursing and social work research, hardly qualify as science but they entail and need disciplined inquiry.

In our view, a focus on scientific standards in the review processes distracted attention from an equally pressing concern. Much writing (eg Rein, 1976; Rein and White, 1977; Caplan et al., 1975; Weiss, 1977; Kogan and Henkel, 2000) in policy research has been concerned not so much with scientific standards as with questions about the contribution that scientists can make to policy problems and the capacity of policy-makers to use scientific knowledge.

This suggests that researchers wishing to contribute to policy problems must enhance their understanding of those problems and the way in which policy-makers, administrators and practitioners address them. Some of the work assessed in the Chief Scientist's visits that tested the definitions of science can be seen as a response to these preoccupations. Many people believe that the primary problem for policy-relevant research lies in the understanding of the relationship between science and policy.

The concept of a unified scientific community is an illusion. As we saw in Chapter 8, research units work within different epistemologies and their networks are multiple. Scientific advisers vary in their views of the relationship between science and policy and of the means by which good policy-relevant research is best promoted and evaluated. Their own judgements stand in complex relationship with those of policy-makers in the Department who, in turn, are often proxy customers for other members of the research unit networks, such as practitioners, administrators and consumers. This case study presents a picture of multiple authorities and multiple knowledge systems that provides an argument for alternative approaches to evaluation. Such approaches might be based on beliefs in pluralism, greater participation in decision-making and more exchange of knowledge. They might entail interactive modes of evaluation in which criteria and processes were geared to the objectives of units' research and its place on the policy-science continuum. Thus, no single blue-print of evaluative arrangements would be appropriate (Weiss, 1977; Lindblom and Cohen,

1979). The story of this particular form of encounter is given here in detail because it exemplifies many of the major themes already substantiated in this book. It puts under detailed scrutiny the ways in which scientists took a view of the kind of science sponsored by the DHSS. It shows how those often conflicting scientific criteria were then placed under the judgements of policy-makers as to their usefulness and relevance. It suggests that changes in views of science and of its relationship with policy may need to be reflected in structures and processes of evaluation.

Part III Emerging Processes and Roles

Chapter 11: The Functions, Process and Impact of Research Commissioning

OUR EMPIRICAL chapters (4-10) described how structures developed and produced interactions between the main actors. These events are reconstructed in this chapter to display the different functions attributed to research, the processes of promoting it which resulted from those functions, and the evaluations made by some of the main actors of the impact of the research commissioned by the Department. In Chapter 12 we show how the same developments precipitated particular roles and relationships.

The function of research

Some customers took an instrumental view of research and hoped that it would provide data addressed to specific problems. Others hoped that it would enhance the general understanding of problems faced by policy. Research management advocated yet a further range of purposes.

The viewpoint of the customers varied according to their position within the hierarchy and the kind of policies which they administered. Customers mostly wanted data to help them do their work. An internal departmental survey of demands referred to needs 'for information which help the Department's policy-makers to know more about what is happening in the world of policy provision.... a further. . . use is for research [as] a means to policy goals, usually by enabling the Department to get a particular development off the ground for a trial period in a relatively cheap and uncontroversial way, without having to make a long-term commitment.' (IM (80)/1). Some customer views of the function of units emphasised the need for 'sound' research which would help create feasible policies or practice. Whilst one interviewee (ID (80)/18) thought that units should not only answer questions put by the Department, but also ask questions the Department had not thought of, the more general view was that units should answer the policy-makers' questions and their work should be practical. As we have seen, another wanted tighter and clearer contracts with units. One division wanted research that might point towards better resource allocation or help with costing and clinical budgeting. Customers were interested in the assessment of needs and valuation of performance (ID (80)/20).

At the same time, the Department presented a less instrumental view. Customers with a policy analytic remit thought they might seek help outside the Department if they lacked specific expertise or if they wanted a 'non-civil service kind of look at something'. The approach of the outsider could be research- rather than policy-based and concerned with long- rather than short-term issues (ID (81)/2). More basic research could recast thinking about needs and services although such issues might be excluded because of their high political content.

A DHSS contribution to a seminar stated that it 'values its contact with the academic and scientific community not only for the results of specific inquiries carried out on its behalf, but also for the depth of theoretical and historical understanding which is thereby brought to bear on its tasks and the critical but not unfriendly involvement in its business.' (Barnes and Connelly, 1978). A customer at the top of the system implied a more liberal approach to policy formulation and its relationships with research. Observation was relevant and important to policy. Scientists could bring objectivity and independence to establishing factors that effect change as well as to analysing present conditions. The policy-maker must value the research worker as an interpreter of nature, not just as a gatherer of facts. This might open up unexpected discoveries and relationships hitherto unobserved:

> One might venture to say that the requirement for scientific background is most pressing where there is no certainty that the current state of affairs is wrong or about the practicability of a remedy if it were, and yet much turns on the truth of the matter, in terms of human well being or use of major resources. I doubt whether these critical zones are easy to spot, and if they are to be spotted and worked over it can be only by a very close partnership of policy developers and researchers anxious to support them.

These precepts were, it was hoped, translatable into relationships between research and policy development and research and the planning system (Church House meeting, 1979). Policy-makers could be sustained by knowledge of the system including that derived from scientists' observation of, for example, the process and impact of health services organisation. As resources became more constrained, and government attempted to be more devolutionary, this function would be more necessary. Yet, for the most part, such conceptualisation involved collecting data rather than more theoretical work (ID (80)/11). Also, there was a time-limit on usefulness. It was not possible to commission work that would help the problems of five years hence.

As Chapter 7 showed, the Research Liaison Groups became the principal machinery for customers' expression of needs. Yet customers without RLGs could work directly through research management and some were entrepreneurial. A professional division, for example, promoted research on prescribing patterns and also ways of changing doctors' attitudes towards drugs. It also developed research networks of practitioners in interactive research bringing doctors together to compare practice and thus use peer groups to establish causes of changes in behaviour. Detailed customer involvement with research development outside the OCS structure was rare but reflected some feeling that the RLG device was cumbersome and could 'gum up the works' (ID (80)/18). The range of research, therefore, included that directly commissioned through the OCS, or on the agency of customers without RLGs, and research as well - perhaps 50 per cent of it - which did not originate from a customer initiative. As far as researchers in the field were concerned, however, the Department seemed reactive rather than proactive. Our interviews inside and outside the units suggested that, for the most part, researchers devised programmes which the DHSS then either accepted or rejected or modified.

Researchers complained that they could not easily know how the DHSS identified issues for research. In some cases there had been no contact with policy-makers. Some liaison officers were positive and helpful, but others less so. When requests were made the units felt them to be unclear and the Department was unwilling to discuss the work to be done. Researchers were not properly informed on the policy or practice implications of what was wanted. There were conflicts of views between divisions. Where units worked directly within the ambit of an RLG, research might be initiated more co-operatively, but that might be because some areas of study were more easily aligned to some areas of policy than others.

The complaints partly reflected the range of issues upon which researchers' help could be sought. Research on short-range policy concerns might demand easily specified information. But on medium-range issues, the Department must predict a requirement three to five years ahead, although it is virtually impossible to predict policy movements with any certainty. Paradoxically, longer-range work might be easier to commission because the pressure to be realistic is less, although the incentives to do so might also be weaker. Two leading researchers said that whilst they were concerned with action and practice, the Department's view was almost academic. As one put it: 'The DHSS was a kind of electronics firm that put funds into research on fundamental electronics, or on the number of visits that salesmen make, but in no sense [was] concerned with selling products.' (IR (80)/38; IR (80)/37).

The Department's difficulty in specifying needs in ways expected by researchers was sharply exemplified by the case of the Panel on Medical Research whose history we have related. The scientific community expected too much of the Department's ability to specify its needs in bio-medical research, and there is a lesson to be learned from this. Where there was a match of expectations, as with some RLG initiatives, it was because researchers and policy-makers joined forces to identify research needs against policy relevance and feasibility. Government cannot state its research needs from 'cold' and hope that good results will follow. As customers developed different views of their research needs, research management considered how to formulate research priorities. A DHSS research management paper (CSRC (77)/1) referred to 'a responsibility for seeking the greatest amount of useful and effective information . . . for whatever sum of money is available for research.' As we have seen (Chapter 5) it declared that in choosing research priorities three sorts of criteria were available: the service, the thematic (those research objectives that potentially cut through the whole departmental research programme) and the scientific. No decision followed this speculative exercise but the emphasis fell on service research even if thematic studies were always thought desirable and possible.

Like the RLGs, research management thought that documents such as White Papers and Consultative Documents might trigger off the discussion of priorities. One former research manager (ID (80)/5), however, thought researchers should be able to influence customer views about research needs. Research management considered whether resources should go to areas of greatest growth or 'to areas placed under the greatest strain by the absolute or relative loss of resources'. RLGs were established first in areas considered under-researched. As a result, as we have seen, physical handicap greatly increased its share of the research budget whilst that of hospital services fell from 29 per cent to 18 per cent.

Research managers (IM (76)/1) noted that while the RLGs and policy divisions must formulate research strategies, they tended to concentrate on short-term, one-off research projects and on the detailed and graspable. Short-term projects were justified, but they might inhibit the development of people and methodologies. Researchers themselves often preferred deep but limited projects rather than research on more nebulous general problems of concern to policy-makers. Ways had to be found to reconcile short-term policy concerns with the preferences of the research community: RLGs must recognise that planning must not be too specific if researchable projects were to result.

The statement of these cautions reinforce our point that the Department was not monolithic or uncaring or unnoticing of the achievements or the needs as well as the deficiencies of their research contractors. As long as the CSRC and RLG version of Rothschild seemed workable (1972 to 1978) research managers struggled with these issues, until the focus shifted to concerns about scientific accountability.

A study undertaken in 1976/77 by Moss, based on interviews with administrative and professional staff, indicated that the research projects valued most highly were: examinations of how a particular service was currently working (54 per cent); feasibility studies of new ways of meeting a known need (50 per cent); and descriptive accounts of an existing situation (46 per cent) (Moss, 1977). The least preferred were clinical laboratory research which examined the nature or the cause of illnesses or evaluated ways of treating it (21 per cent); and studies designed to compare costs of alternative ways of meeting a known existing need (18 per cent). These findings, it will be noted, are consistent with other research challenging assumptions that administrators prefer studies that merely reinforce and legitimate present ways of working (Weiss and Bucuvalas, 1977).

Commissioning and monitoring

Once initiated, projects were commissioned and then monitored by research management. The relationships between major contractors and customer divisions could reduce research managers' control over the processes of ensuring that competent contracts and budgets were created (ID (81)/4).

There is no reported evidence that the DHSS was other than fair-minded and liberal in the terms that it negotiated. But making contracts gave rise to large issues. The Department usually restricted contracts to three-year periods; more rarely it gave rolling contracts of up to six years. Researchers, however, protested the need for continuity of research activity, as well as for reasonable job security. Such issues as the amount of overhead allowances, of the free time assumed by contractors, and the rules governing the publication of funded research, are important to both sides.

Liaison officers managed the exchange of research findings; annual reports of projects were a minimum means of securing accountability for funds raised. They had to give continuing support to researchers whilst monitoring their performance.

The evaluation of single projects was more variable than that of DHSS-funded research units. OCS sometimes sought comment from external advisers before sending reports to the customer division or the RLGs. At other times it relied on RLG advisers who knew the policy context. Some final reports, again, were sent straight to the customers. In other cases, liaison officers set great store on rigorous checking of the scientific validity of completed work.

The Department made a start on linking the evaluation of reports with the consideration of further research studies and the development of policies and practices (RLG (MI) (77)/33). The variations in actions following evaluation have been well documented by Gordon and Meadows as they analysed the consideration of final reports by RLGs in terms of discussion of policy implications, of research implications, of dissemination and of the initiation of other actions. (Gordon and Meadows, 1981).

Reception and dissemination of research

While almost all researchers could point to some impact of their work, they were less certain of its reception. At a meeting with the Chief Scientist (Church House meeting, op. cit) 'it was felt that researchers often had no way of knowing whether their work influenced policies. In the field of personal social services research, publication and dissemination was [sic] a particularly weak point.' When the newly-appointed Chief Scientist asked unit directors (in 1978) about the use made of their research the replies proved difficult to generalise. But at that time most of the work had not been commissioned at the request of customers because it pre-dated Rothschild. Nor had it necessarily affected DHSS policy - most was more useful to the field.

What does this process look like from within? The Department had promoted research to help it solve problems. How open and active was its reception of findings?

If there was an RLG to receive the research this entailed receipt by the customer. After the customers and others had commented, the liaison officer communicated views to the researchers.

While many researchers felt their work was ignored, or they did not know whether it was ignored or not, the Department's liaison officers committed time to evaluating and circulating findings to a wide departmental audience for critique and noting. Most letters researchers received from liaison officers attempted conscientiously to collate the criticisms made. Yet if they discharged the duty to make some kind of response, the letters were not always of the kind for which researchers hoped from a sponsoring body. It may always be difficult for the Department to say 'this is not only interesting but will affect our policy in the following ways', if only because research may affect only particular issues or may have unpredictable long-term effects.

Our reading of Department files is confirmed by the more detailed study made by Gordon and Meadows (op. cit). While 51 per cent only of the researchers interviewed claimed to have received feedback from the Department, all of the liaison officers interviewed could demonstrate that they gave feedback on almost all reports submitted with fairly detailed comments in their communications to the majority of researchers. Yet not all was transmitted. In one case, a piece of work regarded highly by the Department was thought by a unit to be totally ignored by it (IR (80)/38). In another case, a piece of work which the Department had good reason to think poor was ignored although it related to specific policy actions at the time (IM (78)/1). Perhaps the Department could have said frankly to the researchers, 'This is your view on an important issue, but it is not going to be ours.'

A great deal of thought was given to dissemination (PMR (76)/25). If researchers could be counted upon to disseminate, research management nevertheless felt that policy divisions must be the prime movers in taking up the fruits of research. Proposals were put, in fact, to the CSRC in March 1977 (CSRC (77)/14). The research managers hoped the policy divisions would commission a scan of the long-term results to enable them to develop criteria by which to judge newly completed research. They should take into account the academic structures and disciplines underlying the research, not only as an aid to validation but also as a broader context for judgement. Researchers were already required to state the policy implications of their work and in some cases had done so. Research management felt that they might formalise this process into 'debriefing sessions' halfway through the research as well as at the end to include critique by a scientific adviser and a member of the policy division who should provide a view of the results for the RLGs. This device had, in fact, been used. It might be possible occasionally to commission an external assessment of findings on both scientific validity and policy implications. Results might be disseminated through face-to-face encounters, through notes to field authorities and training bodies, by the publication of highlights, even by the commissioning of scientific journalists to prepare their versions of research reports. It might be possible to recruit 'propagandists' to spread research findings and help the field authorities put them into practice. These and other proposals were carefully worked out and to some extent found their way into practice.

Some RLGs held one-day conferences with researchers. Two RLGs each held specialist seminars on the use of case registers. Another two RLGs held conferences on completed research to disseminate and enhance appreciation of work already done. Other RLGs had meetings at which research, either prospective or completed, was considered with researchers in their fields. These efforts fell short, however, of systematic joint attempts to work through the consequences of the research commissioned.

The Department was certainly nothing other than liberal on the dissemination of research findings. Its research programme is an open one as indicated by answers to parliamentary questions throughout the 1960s, and an answer given by Sir Keith Joseph in 1971:

> Reports to my Department are required in respect of all research commissioned. In general, researchers are also free and expected, but not obliged, to publish the results themselves. The draft of the publication is submitted to the Department for preview, but, subject to the exclusion of references which might lead to the identification of individuals who have co-operated in the investigation, the author is free to reject or accept any comments made.

The DHSS was hostile to a Treasury attempt to impose the restrictions of Crown copyright on publication (Gordon and Meadows, op. cit).

If, then, the DHSS, within its limited resources, attempted to commission, evaluate and disseminate the findings of research, what impact did this have upon the Department? Here we must face two difficulties. We shall soon give some examples of quite direct impacts on the DHSS but there is a substantial literature to the effect that research has little linear or direct impact on policy but instead works through 'diffusion' or 'percolation' or 'illumination'. Our second problem concerns the time-span over which the case study was taken. By the time our main field work on the RLGs was completed, hardly any work commissioned within the Rothschild era had come through as completed research or in the form of research reports. This being so, much of the evidence relates to work that was already in the pipeline. In brief, however, customers became increasingly clear that certain kinds of research had impact on policy, and that the processes engendered by the RLGs helped them clarify their policy fields.

The evidence of two customers followed the Weiss assumptions. Research makes the client group think. As they absorb research it influences them, if unobtrusively, and cumulatively. Research can help them in discussion with pressure groups. Research both raises questions and reinforces the questions that customers formulate. Even when research does not confirm policy-makers' expectations it makes them ask why they do not agree with the research. Research challenged a political view on, for example, the maintenance of local hospitals for the adequate treatment of emergencies (ID (80)/19).

Another senior interviewee (ID (80)/11) pondered on the reasons for the Department's difficulty in absorbing the results of research. The research might be unclear in its messages and then only suggest the need for more research. Policy-makers might be too busy working on too wide a front. Research might have focused on too small a proportion of the picture to be of direct use. Timescale may be important. Some policies change very quickly but others as, for example, concepts of good community provision as promoted by RLG work are not likely to change for a very long while.

The perceptions of individual interviewees were reinforced by a departmental exercise on the usefulness of what had been commissioned (IM (80)/1). This was undertaken as the Department faced challenges on the quality of its research commissioning in the late 1970s.

Its analysis was based on over 200 examples of 'usable' research provided by policy-makers and research managers over the whole extent of the R & D programme. It was noted that there were several ways in which the programme might bear fruit: through individual projects; by the accumulation of a 'critical mass'; by providing information which enabled service providers to have an increasing understanding of the basis of their work; and by reflecting the general climate in which the Department's work was carried out. There could rarely be a one-to-one relationship between a single research report and a policy or service development;

the Department had not, in fact, supported that sort of research. Sometimes, a project demonstrated that a new development was not necessary and this was 'every bit as useful as a more positive result'.

The Department's evidence was that customers and research managers could detect useful elements in the commissioned projects. This might have been a projection of parental expectations of their progeny, but the evidence convinces by being detailed and unsensational. Important facts about childcare systems might be disclosed and thus give clues on how practice might be developed. Comparative costs of particular kinds of social treatment relationships between services in the same area, surveys of what field authorities do with particular client groups, were among the studies. Evaluation of particular forms of hospital service, important negative evidence on the impossibility of predicting a particular form of disablement by measurement of blood pressure, were others. One survey of a form of social disability 'provided a defence for the DHSS against those who wished for large extensions of services to this group.' Another study provided the Department with options over a major administrative and financial issue. The list is long; the examples quoted show how research was thought to help policy-makers and practitioners with the bread and butter of their daily work although some went farther. One study 'together with several previous studies' by the same researcher was said 'to have been exhaustively used in policy formulation by the Department'. Another project financed by the Department was now 'an extremely common base point in all social work research' concerned with the development of particular review systems.

But discussions in RLGs of completed research reports indicate the timescale and the collaborative mode required to enable research positively to contribute to policy. Research on a particular form of care for the elderly raised questions about 'the whole concept of this form of provision' (RLG (SHB) (M) (79)/1). Research on factors affecting perinatal mortality was found to have produced important policy indicators, but more fine-grained research would be needed if any clear guidance on policy priorities were to emerge (RLG (CH) (M) 27 Nov.1979).

RLG work on research needs in intermediate treatment vividly illustrates the gulf between the broad questions of policy-makers 'What is the effect of intermediate treatment on a young person's subsequent history of offending ... what effect does intermediate treatment have on a juvenile as a person?' (RLG (CH) (80)/13) and the perspective of researchers. The paper commissioned by the Department to identify research potential in the area pinpointed the formidable conceptual and methodological problems of the field. The inter-related strategy proposed of 'outcome, processual and economic research' would still require policy-makers wanting answers to cast their bread upon the waters. The potential of research for problem elaboration was clear; but for problem solution highly uncertain. In such fields the interactive model of research impact is the only convincing one.

In describing the different elements of the commissioning process, we have given an account of the direct and indirect impacts of research on policy-makers. The case of the DHSS confirms Weiss and Bucuvalas's observation that policy-makers do not commission research simply to reinforce existing policies and practices although there must be some governments, in some places, and at some times, that do so. If research can affect specific policy perceptions and developments this cannot be generalised for all commissioned research; nor can we, from our study, confirm the more general percolation effect of research although this was commonly avowed by customers and research managers. Nor was there strong evidence of links between the Department as proxy customer and the field authorities as implementers of research findings, although some research commissioned by the DHSS had been replicated by other researchers and field authorities too had been involved. That did not add up to widespread dissemination and implementation of findings. The DHSS might have been inhibited in giving a lead in such activities both by the failure to build adequate machinery for connection and by the intrinsic difficulties of transferring conceptualisations of recorded practice into work in the field. Yet the instrumental value of commissioned research is also plain from the examples given above.

Chapter 12: Emerging Roles

NEW ACTIVITIES produce new structures. They can also precipitate new roles and change existing ones. As a result of the experiences recorded in the previous chapters, roles which had been latent as well as entirely new types of roles began to emerge in the DHSS. The two most important were those of *customer* and of *broker.*

Acting as a *customer* proved to be a far more difficult task than contemplated by Rothschild. The processes of identification and commissioning made it necessary to negotiate several points of view at any one stage. The simplicity of the Rothschild customer-contractor principle became modified in new, or reformulated, functions and roles to go with them. In particular, we note *'brokerage'* which appears most obviously in the roles of the Chief Scientist and of research managers acting as liaison officers between research and the policy divisions.

The customer role

The customer role was played by the policy divisions acting through the RLGs or through research management. Our account of the working of RLGs has indicated the potential for variation of that role. Research commissioning is a small part of the work of policy divisions and research is a variable input to policy. A minimum role description, however, of the research customer would include the following functions: to scrutinise areas of policy and practice in order to identify lacunae in factual data, service evaluation or conceptualisation; to determine whether the gaps can be filled better by researchers than by the DHSS itself; to ensure that practitioners and field authorities help specify research needs; to state needs well enough to enable research management to find appropriate researchers; to receive, evaluate and assimilate research findings into policy-making; to promote dissemination of the implications (if not the substance) of the commissioned research. This is, of course, an idealised account of what happened.

Government departments may commission research for their own purposes or take on a secondary or proxy customer role in which they commission research for the benefit of the field. The DHSS played both the primary and the secondary customer role. Not all government departments do so but in health and personal social services, however, the Department's role was partly to set frameworks within which others work. It might then furnish information useful to service providers. That gave the Department a secondary role. And it made the DHSS a proxy customer for the field authorities and practitioners who might directly benefit from research commissioned by the Department.

Central administrators find it difficult to be customers in the primary mode. Involving researchers in strategic and 'in-house' issues is politically difficult and not part of the British central government tradition. Nor is social science's contribution to policy formulation well established. Whilst some research relevant to central policy-making is better undertaken externally (research on policy for social security, or the effects of the Resource Allocation Working Party (RAWP) are obvious examples), other issues might best be tackled through policy analysis which administrators may think they can undertake as competently as external researchers. The secondary mode entails empathising with the needs of the many fields with which the Department is concerned. This requires a good knowledge of field practice and might lead to search for data and concepts which do not easily fit existing policies or patterns of practice. The primary mode, we suspect, would require researchers to encounter policy-makers in their more authoritative, controlling and convergent styles, whilst the secondary mode responds to developmental and more open styles in government.

The customers worked predominantly in the secondary or proxy mode, identifying research questions more important to the field than to the DHSS, in the areas of client need, of professional practice and of the organisation and management of services. A strategy statement exemplified this point:

> The RLG's central aim is to sponsor research whose results will help practitioners in the health and personal social services to give the best possible services to mentally handicapped people, will help those responsible for planning these services to make informed choices, and may also be of direct assistance to families who both receive and provide service. (DHSS Handbook, 1979)

If, then, the customer role was multiple, how did it work? First, the customers were primarily those in the Service Development Group who were concerned with client groups rather than with the field authorities. Some such customers (ID (75)/1) regretted their lack of direct contact with field authorities because it was the regional directorates, rather than they, which had access to them.

Increasingly the policy divisions felt that acting as customers helped policy development. Positive feelings towards the RLG operation grew during our period of observation. Presenting a paper to an RLG meant taking a view hitherto inexplicit.

Indeed, the RLG might force the Department to take a position in order to seek advice (ID (76)/11, 15). Sometimes quite low-level work was needed in the first instance; descriptive pieces about how social work was practised, or better assessments of existing techniques and practice in the field which seemed more related to development than, research. Research managers felt that customers often demanded, indeed, too much detail from research. Research management preferred a more free-flowing approach; no proposal would meet the exact requirements of customers. Better that the researcher had a clear understanding of the needs of the Department and sought adequate critique of methods (ID (76)/20).

If research management was willing to sponsor a wide range of approaches, Gordon and Meadows (1981) noted how the formal and schematic nature of the prescribed procedures contrasted with the individualistic and interactive behaviour of its lead officers. As we observed it, contrasting views were aired. One statement assumed that the customer could establish clear requirements for research to be undertaken by a systematic process of discovery and quantification and embodied a schematic procedure through which research might pass before it could be commissioned. A senior professional associated with research management noted, however, that 'The administrator tends to believe that research can be managed more closely and directly than does the professional.' A reference to the divide between the customer and the research manager which might be applied, too, to divisions within research management. There might be, it was conceded, more pressure on administrators. But administrators

> are not so able to use groups and to understand that they take time to get working. Committees are started, or a system is started, and it is expected to work in exactly the way in which it was set up . . . . it is not thought possible to try something out in small way and to build up from that. . . . The more important point is to see that when research proposals are being formulated the questions or hypotheses are as broadly based as is possible . . . [so that] although a narrow focus is frequently wise and necessary, we avoid the situation where . . . the projects take no note of related issues. (ID (75)/4)

In thinking about ways of getting the process started, research management, administrators as well as professionals, became aware of the somewhat specific objectives followed by the customers. One administrator noted that RLGs would follow 'a fairly circumscribed path' perhaps because professionals did not assert their views sufficiently within the Department. Objectives were built in too quickly because the policy divisions must determine them before RLG or other meetings. Yet the

RLGs that had done most might be the more restrictively led while those whose work is more diffuse might produce more open outcomes.

Styles of RLG chairmanship ranged from the downright dictatorial to the virtually ineffectual:

> [the] typical RLG Chairman, if any such existed, might be portrayed as functioning with a certain basic good will and willingness to give science a chance to see what it can do, tempered by an administrator's ingrained scepticism and a resolve not to let science dictate its own terms. At its best, in encounter poised to a strong and enthusiastic scientific input, this can contribute to a healthy balance of forces; where this counterpoise is lacking, it can lead to an RLG spending years hovering around the periphery of its subject without ever really getting to grips with it. (RP (M) (79)/2)

Throughout, the policy divisions, although increasingly willing to act through RLGs, were concerned about the lack of resources with which to play the customer role. The difficulties of finding time was a repeated theme (MG (79)/3), significant because of the general assumption underlying the DHSS's application of Rothschild that scientific advice should usually be called upon after a policy initiative had been taken (CSM (79)/6). Staff, too, turned over so quickly that it made it difficult to build up a working relationship with advisers. And working with researchers under the Rothschild precept was often 'an unfamiliar process' (Korman, CSRC (76)/6, Addendum). The Chief Scientist felt that 'in the absence of a stable population on either side the role of research management was to relate one to the other.' (DHSS MG (78)/9). The divisions seemed to be dealing with problems rather than with devising over-arching policies. But policies had to be evaluated before a research problem could be identified. Policy divisions had information which was not organised to be converted to researchable issues. It was not easy to draw on the resources of other parts of the DHSS, or often of the field, so information might be brought together at the RLG meeting rather than before it.

If customers increasingly valued RLG contributions to crystallising policy objectives, this did not justify the customer-contractor principle in its original and simple form. It assumed a great deal about the relative power of researchers and government. The principle was, in fact, a glib metaphor that concealed the weaknesses of one side of the relationship.

It assumed that functions could be allocated between science and government. It assumed, too, that there was parity of power between government and research; but the institutions and individual researchers drew their resources from the customer department. There was no free-market relationship between a customer with money to spend and a seller who could sell his wares elsewhere if the price offered was not right. Further, the metaphor of the 'customer' did not take account of what was being purchased. Some of the work commissioned might be 'short order' analyses or information gathering; but the Department had long attempted, well before the Rothschild Report was implemented, to develop a research capability with the necessary freedom to offer long-range contributions to problem solving.

The metaphor of the customer implied someone who knew what he wanted. But research should help customers to raise and to redefine questions. The metaphor of the contractor, too, implied production units ready to work with established technologies at their disposal. Many of the methodologies needed for tackling the issues faced by the Department were at an embryonic stage, particularly in the areas of service evaluation research.

The Chief Scientist

The first two Chief Scientists were scientific advisers leading an advisory system. In 1978, however, the model was radically changed when the third incumbent of the post also became the DHSS's research manager. There were, at the same time, important developments in the perception of what would constitute good commissioned research.

This extended role comprised three sets of functions: The Chief Scientist remained the Department's adviser on science; he became managerially responsible for its commissioning; he also became, but this was never fully explicated, the principal broker between the policy needs and scientific development within the Department. Although conceptually separate, however, the fulfilment of the three functions was strongly interdependent.

As chief scientific adviser to the Department, the Chief Scientist adopted a firm position on the kind of science which the DHSS should commission which then strongly influenced his view of the way in which the DHSS should manage science. He was concerned most particularly to advance the concepts of 'scientific accountability' and scientific excellence. To this end he advocated a reduced role for the RLGs at a time when the customers themselves enjoyed fewer resources so that they needed more rather than less support from the OCS in their functioning as customers (CSRC (79)/2). The Chief Scientist, from 1978, advanced a policy predicated on the assumption that the DHSS should put its resources into establishing high quality and independent scientific bases. From this assumption flowed the policy of returning the transferred funds to the MRC and of appraising the DHSS-funded units by criteria more familiar to those engaged in MRC-style science than in DHSS policy-related research. The discriminating edge, therefore, of the Chief Scientist's judgement was applied more to scientific advance than to the advance of policy and practice through the medium of disciplined inquiry.

The advisory role, therefore, directly affected the Chief Scientist's work as the Department's research manager. During his time, the research units were substantially reduced in both number and funding and a reciprocal increase in the allocation to the MRC, which was then to favour health services research more than hitherto, took place. In the same cause, the visitations to units were administered on a systematic basis, and refereeing of all research proposals became tougher.

The third role, thoroughly permeated as it was by the advisory and managerial functions, was that of broker. The Chief Scientist was a leading example of brokerage in government. The Chief Scientist to the DHSS had, in fact, several brokerage roles. He had to act as a principal intermediary with the research councils, and that was, in fact, the main task to which he applied himself. But it was also implicit in his role that he should act as a broker between policy and science, and this was a set of functions never made explicit or fully legitimised by the Department or built up, on his own volition, by any of the Chief Scientists. Although the Chief Scientist became a member of the Department's HPSS Strategy Committee and the Department's Research Strategy Committee, this function was more advisory than that of a broker or intermediary. Attempts to formulate strategic scientific policies, as had been made by the CSRC and one of its panels, were not reinforced by the Chief Scientist's membership of the committees.

The brokerage function assumes that policy needs can be transmuted into research questions. It implies a reflexive view of policy-related research in which thinking about policy and thinking about policy-related research are not separable procedures. It implies an entrepreneurial role in which untidy and messy policy issues are analysed for their potential for disciplined inquiry. It involves the Chief Scientist or any other broker in getting into the skin of the policy system and empathetically drawing out its problems. It assumes that the starting-point for good policy-related science is not in fact science but the policy problems it attacks. It is, therefore, somewhat in conflict with the policy of establishing 'good' science in independent research bases, perhaps financed by the research councils, rather than in DHSS-funded units and projects.

The Chief Scientist as broker and mediator must, therefore, face several ways. The contacts with research councils, and the kind of science which they can provide for policy, constitute a central task. But other contacts are equally important. The DHSS cannot face its policy problems alone but has to use data and concepts derived from sources from outside and then bring them together into a network of knowledge and action. In so doing, it could exploit several sources of data for analysis. Apart from commissioned research, there are the connections to be made with the professional groups both within the regions and districts and within the DHSS. A further capacity is the Policy Strategy Unit. It was not clear to us, however, that any of the Chief Scientists were able to make the connections between commissioning and receiving research, and the other forms of analysis and planning of which the DHSS was the centre. Moreover, knowledge of developing practice in the field could be the starting-point of research. Without strongly-grounded knowledge of developing needs and practice, policy-related science is an empty thing.

The triple role, as we have described it, places an enormous burden upon the Chief Scientist. It requires somebody selected as an eminent scientific specialist to take on some of the characteristics of the generalist. As we have seen in Chapter 3, the specialist manager has an ambiguous role in government where generalist administrators coordinate and adjudicate between different viewpoints. The art of central government is largely that of bringing together perceptions ranging from the political, generalist and value led to the specific, technical and factual (Brown, 1970). The Chief Scientist as adviser can be a specialist, although he must draw on expertise well beyond his own. The Chief Scientist as manager has to work largely on criteria which are those of the generalist because his concerns are primarily those of policy rather than of science. As broker, he is deep into the generalist mode.

Moreover, apart from the enormous range of tasks that the triple role creates, it also requires multiple modes of behaviour. A Chief Scientist acting as a manager exercises authority within a formal structure. He is no less a hierarch than is an administrative officer. Nor is it obvious that team work or collegiality are stronger among professional and specialist groups than among generalist administrators if they are concerned with administering government business. Research management must, therefore, rest on organisation whilst at the same time remaining reputable for its expertise or its access to experts. It forms part of a system which has to be coordinated and within that larger system specialist managers have to ensure congruence and, in so doing, become 'generalist-specialist'. Managerial behaviour is conditioned and not only by training and professional allegiance but also by task.

In our view, the enormous complexity of the Chief Scientist's job was not adequately thought out. Nor was the full potential of the role accepted by administrators who saw themselves as taking the lead role in policy formulation. Policy-makers in Britain did not welcome policy analysis or other forms of academically-based policy critique and development. This was made all the more difficult by the fact that the Chief Scientist had tenure for no more than three years. Obviously, a scientist entering government for too long a period would lose his expert status and therefore his credibility with the scientific community. Ten years would have been too long; three years was too short.

We have assumed that this complex role required far more explicit and firm relationships with the policy system within the DHSS and with the fields of policy and practice outside the centre. In effect, however, the Chief Scientist's main connections were with the advisers whom he appointed and with his own subordinates in research management, later to be called the Office of the Chief Scientist.

Advisers

By 1977, when the three intermediate boards had been disbanded, there were 90 advisers serving on all levels of committees and groups advising the Chief Scientist. The disciplines represented were: social administration, social studies and social work; economics; nursing; social and community medicine; psychiatry; sociology, social anthropology and criminology; medicine, including pathology and pharmacology; engineering; psychology; computer science; biology and physics.

While the Chief Scientist had a multiple brokerage task his advisers had more restricted functions. Yet they, too, encountered ambiguities in matching their expertise to the Department's needs. Research management was concerned that advisers might be thought to lose their independence if they were too helpful to the Department. The scientific advisers themselves, when considering the point, felt that

that had not happened (Korman, CSRC (76)/16, Addendum). There was a feeling that communications were improving even though there were perpetual uncertainties.

Some advisers felt frustrated by what they saw to be both a restricted and an overburdened function. DHSS officials, too, considered how to strengthen scientific impact and, before restrictions set in, whether to recruit scientists on a part-time or temporary basis instead of for a few meetings a year so that they could play a more effective part without losing external academic status (RP (76)/20). Whether scientists seconded in this way could operate in the complex research management network would have to be judged in the light of experience. But the analogous secondments of economists to the Treasury are well established and apparently successful.

Scientific advisers operated on the boundary between scientific judgement and policy identification. That relationship worked best when the policy-maker decided what he needed and the adviser decided what would constitute good science. But there were many examples of both pondering on the best area in which to instigate action. This process had to be largely intuitive and interactive once it was realised that ambitiously systematic methods of identifying researchable areas were too extravagant and remote from reality.

Because advisers need to move between research and policy, research managers attempted to clarify their roles. They were both critical and supportive of them. They thought that advisers took a somewhat naive view of the operation of government, and did not understand the structure of the Department or the place of research management within it. Advisers' anxieties about how to perform their functions were accentuated by lack of understanding of how power was dispersed through the Department and the research commissioning system. It is not, however, clear what role was envisaged for advisers who might know their way about the Department.

Some advisers were thought largely to represent their own views, to advance particular interests or viewpoints, and to speak as if customers rather than scientists. They could hardly represent a whole range of disciplines nor yet easily come to terms with policy issues. No adviser was likely to be able to provide all that would be needed by an RLG working at full blast. Whilst they might be able to state research priorities, they could not necessarily relate them to service issues. Some were concerned with analysing the policy issues and others with vetting applications. The

Department felt it needed people who could help research planning through knowing what was needed for the supervision and implementation of research. In general, research management felt it was difficult to appoint advisers who were both good scientists and good advisers (ibid).

At the same time members of the CSRC who advised at different levels believed that many of the difficulties were overcome. So far from becoming captive to the DHSS they were often not used enough. They preserved their independence. In spite of these criticisms on both sides, however, the observer concludes that, given the enormous complication of the system, the burdens placed on both sides, the Department got good value from its advisers and reciprocally many advisers, if sometimes frustrated, gained entry to policy issues and a chance to have a say about them.

The role of liaison officers

As with the Chief Scientist, the role of the research managers who acted as liaison officers with units and programmes and projects reflected the widening concerns of both science and government. The demand became more insistent for forms of intermediary analysis, for network building and maintenance, and for communicating complex messages from one part of multiple systems to others, although the status of the resultant roles remained uncertain in comparison with that of line authority.

Research management comprised administrators responsible for the financial and contractual aspects of commissioning, more senior administrators concerned with general research policy, and professional research managers who referred both to the Chief Scientist and to the professional heads of the nursing, medical and social work groups. In 1980 the equivalent of 10.5 full-time officers drawn from the medical, social work and nursing divisions worked in research management.

Our picture of the liaison officer is composite and idealised, drawn from observation at meetings and from noting their work as seen on the official files. First, liaison officers had authority to decide on some of the requests for resources. That function is conceptually separate from the brokerage role in which they mediated research findings to policy-makers and fed back the needs and reactions of policy-makers to researchers. As brokers they had to understand enough of current trends and methodology of research to form primary judgements enabling more expert scientific assessment to be brought in. They had to be familiar with the principal researchers in fields that might be extremely wide. They had to judge what problems might be susceptible to which kinds of research. They needed to know the limits of tolerance of practitioners who might be the subjects of the research. At the same time, they had to understand departmental procedures and policy problems well enough to help customers to decide on commissioning.

Liaison officers had a particularly sensitive task in advising on the choices of external expertise. For academics may be, in Isaiah Berlin's phrase, hedgehogs (who know one big thing) or foxes (who know many small things) (Berlin, 1953). Advice, too, on the quality of prospective or completed research had to be marshalled with reasonably disciplined procedures. Finding the right people required an extensive knowledge of the academic network and sensitivity in matching evaluators to the research being evaluated. Once advice was sought it had to be used critically and positively.

As liaison officers worked with both policy-makers and researchers, they had to interpret research findings and ensure a proper hearing of them without becoming captive to the researchers if their judgements were to be taken seriously within the Department. Their judgements might be questioned by those who are more expert, if more narrowly focused.

Managers liaising with research had to draw close to the researchers and encourage them in work from which the DHSS hoped to get useful returns. Yet they moved from brokerage roles in order to monitor performance. Both parties knew that liaison officers' judgements counted in decisions on future funding. The liaison officer might work with academics who had moved out of their scientific peer group and put themselves out on a limb to help policy and practice. The liaison officer had to be sensitive to these vulnerabilities but not unduly influenced by them.

Liaison officers were concerned with more than 'their' individual projects or units. Whilst ensuring equity to researchers they had to safeguard the Department's need for flexible research capacities. Issues spanning the interests of more than one RLG or customer division had to be related to each other. This needed confident work in identifying linkages not always evident to either the customer or the researchers. Consider, for example, how the measurement of disability might be treated in such a network. The concept can be generalised but its applications affect the administration of services in many policy zones. The liaison officer working on an individual project ought to link it with others developing across the range of DHSS interests.

The research manager had also to help the scientific community to sustain its networks. This meant dropping the authority of the Department whilst participating in the world of conferences and seminars. The broker was thus led constantly to change roles whilst crossing established boundaries.

At every internal DHSS boundary there are mediators and boundary crossers; a flowchart of research decision-making might show different forms of brokerage growing up with the different policy-making, planning and professional groups within the Department.

Customers outside the DHSS

The DHSS acted as a customer both on its own behalf and on behalf of the field authorities. But field authorities were customers for research in their own right. Most social service departments in our field study (1975) employed their own research staff. Health authorities undertook research from outside research centres. Our interviews with thirteen social service department research staff and with three officers of regional health authorities showed that their research was mainly concerned with service needs. Such projects focused on basic information on the characteristics of the population being served, or the services provided by the authority itself, and the activities and plans of other locally provided services. Attention was also given to evaluative studies: how well operations were being performed; criteria for admission to different facilities; alternatives to residential care; and especially on the social services side, the effectiveness of aids and specific services such as meals-on-wheels and home helps. There was work on the interaction of different services, stimulated in part by corporate management approaches within local authorities, and various joint planning exercises carried out by health and local authority services. In both health and social service authorities, research enabled authorities to examine and plan services on a long-term basis. Thus, the operational aspects of research at the field level involved defining project needs, monitoring developmental work, operational activities, professional standards, and the efficiency of the organisation; providing information for planning and policy formation; determining priorities; and bringing client needs into juxtaposition with political pressures.

There was no systematic connection between central and local commissioning of research. The DHSS regional divisions concerned with health and the staff of the social work service liaised with field authorities. Views on the efficacy of this contact were mixed: some departmental officers felt they were kept well informed. Others agreed with members of the field authorities, who thought otherwise. After a report critical of DHSS working was published (Three Chairmen's Report, 1976) 'representatives' of the NHS and of social service departments became members of research liaison groups. Two RLGs (for the elderly and mental handicap) gained such representatives early in 1978. On the experience of a small number of meetings, this seemed unlikely to have a noticeable impact. It was not clear how an area nursing officer, area medical officer or a director of social services could effectively 'represent' the wide variety of conditions which existed in these services. There was no mechanism by which the field authorities could choose their representatives or receive reports from them. In our view, this kind of representation cannot substitute for the wide contacts needed to cover experience in the field, nor the exercise of the responsibility of the DHSS to disseminate research results to the field authorities.

Apart from formal mechanisms some *ad hoc* arrangements linking field authorities with the Department were made. The Nursing Education Sub-Group of the Nursing RLG held a conference for directors of nursing education when they could express priorities for research. The RLG for the elderly held a similar kind of conference on future research needs following the publication of the OPCS *Study of the Elderly at Home.* By 1980 these, however, were the only conferences held for non-researchers in the field. Members of the field authorities consulted had been generally unaware of the organisation and content of the DHSS research programme.

Acting with the customers in the field had proved difficult for the Department. It did not exercise direct responsibility over field authorities and had to work hard to bring developments in the field and perceptions of authorities' problems and needs into national policy-making.

The kaleidoscope of expectations, processes and impacts of research commissioning and the precipitation of new or reformulated roles following the

Chapter 13: Policy after Rothschild and Generalisations

IN THIS final chapter, we first consider the extent to which policies and structures have changed in the period since our initial (1983) account of the implementation of the Rothschild Report (1972). We then see what generalisations can be derived from the whole thirty years history of inter-institutional working.

The analysis thus far has been primarily concerned with the possibilities of collaborative efforts between government and science, at a time of bold institutional experimentation. In exploring these it has focused upon two kinds of relationship, the institutional and epistemological, and found them to be not separable. We suggested in our first chapter that the reason why it is so difficult for central government and the world of disciplined inquiry to collaborate might be found in the complexities of the systems involved, in the workings of exchange and power in the relationship between them, and in the degrees of compatibility between the cognitive structures and outputs of science and the needs and constraints of government in its various modes of activity. Although we have concerned ourselves with one British government department only, and over a certain period of time, our area of study yields generalisations which we believe to be important to policy-makers and students of policy-making at large.

Developments over the last 20 years

Of the range of concepts from the initial edition that are relevant when analysing the procession of changes, the most significant is probably the concept of the customer for the research. We will first view the institutional history through that perspective. Its changing nature can be considered in three phases. First, from the early 1980s to 1991 there was much discussion about how to meet the needs of the NHS along with the Department of Health (DH) itself as a prime customer for the research. Then the period from 1991 saw the establishment of a new NHS R&D Strategy with Michael Peckham appointed to the new post of Director of R&D. This was seen as perhaps the first comprehensive attempt in any country to develop a national R&D infrastructure for health care (Peckham, 1991; Black, 1997). The position in the first years of the 21st century embodies some continuities from previous phases but also many changes, at least some of which are driven by the continuing search for a structure that will meet the needs of customers. The very definition of the concept of the customer for the health R&D system becomes more complex as greater attention is given to the patients as the ultimate customers (Harrison and New, 2002). The wider public engagement in NHS research takes many forms, including involvement in decisions about agenda setting (Oliver et al., 2004; O'Donnell and Entwistle, 2004).

From Rothschild to Peckham

We have noted (Chapter 5) that, partly at the instigation of the then Chief Scientist, several aspects of the Rothschild reforms had been dismantled by the early 1980s: funds had been returned to the control of the MRC and the detailed committee structure created to fulfil the customer role in the department had been slimmed down. Other aspects of the Rothschild system continued, however, including the Research Liaison Group structure. During the 1980s there were some developments; for example, in 1986 the MRC's Health Services Research Panel was reconstituted as a Health Services Research Committee with some access to funds of its own (MRC, 1988). In the late 1980s the department established a Health Technology Assessment Co-ordinating Group (DHSS, 1988). The new approach to the management of public services, described in Chapter 3, began to impinge on the research system in the UK, as it eventually has elsewhere (Frederiksen, et al., 2003), through the focus on value for money and the economic value stemming from research. In 1987 the Health Department's Research Management Division initiated a study that identified ways of improving and assessing the use and dissemination of research (Richardson et al., 1990). Also in the late 1980s the remit of the department itself narrowed as responsibility for social security passed to a new department.

There was quite widespread feeling, however, that the R&D system was not serving the needs of the NHS: 'the importance of identifying the elements of a *well-integrated system* for the functions of intelligence gathering, commissioning and funding in the realm of health services research is still not clearly understood' (McLachan, 1985, p.4). The House of Lords Select Committee on Science and Technology examined priorities in medical research and highlighted the importance of the customer role:

> The NHS is inextricably involved with medical research, yet the administrative remoteness of medical research from the NHS is a source of weakness to both sides. No research system can function efficiently when the principal customer for research (the NHS) has so small a direct input into the initiation of research programmes. (House of Lords, 1988, para 9.15)

The committee went on to identify two distinct customers: 'The DHSS and the NHS both require research programmes but these will be different in scale and kind. There is a clear distinction between the needs of ministerial policy and of NHS research.' (House of Lords, 1988, para 9.16). We noted earlier the difficulties faced by the DHSS in trying to play the role of proxy customer for the field. The House of Lords committee also questioned the MRC's attitude towards serving the needs of the NHS, highlighting its statement that it justified what it did by reference to its own internal criteria (House of Lords, 1988, vol 11, col 1900). The committee made a series of proposals aimed at ensuring that there was an R&D Programme that was geared to meeting the needs of the NHS.

The immediate reaction from many commentators was that the correct issues had been identified but that the precise proposals from the Select Committee for a National Health Research Authority were 'a dead duck'. 'Perhaps what is needed instead is a national health research policy.' (Smith, 1988, p805). Such an idea, it was argued, would be in line with WHO proposals for countries to have a research policy that would grow out of the national health policies. These, in turn, should be based on the targets for achieving the Health for All strategy (Smith, 1988). Substantial reforms were introduced into the DH's R&D system in the 1990s.

The NHS R&D Programme created in 1991

Whilst the exact approach to reform proposed by the Select Committee was not adopted, the scientific elite who constituted the committee had exerted considerable influence. This parliamentary body which included some of the country's leading basic scientists strongly supported an increase in expenditure in modes of research that traditionally did not have much support, in particular public health and operational research - by which they meant, largely, health services research (HSR). This was seen as a significant boost to these areas of research (Black, 1997) and provides a contrast with the hostility the scientific elite had displayed towards the Rothschild reforms. On this occasion, issues of structure and power were less involved and it was not proposed that extra resources be taken from the MRC. Instead, Peckham soon secured a promise that new resources should be sought through the public expenditure process so that the total should rise from an estimated 0.9% of the NHS budget to a target of 1.5% by 1996 (Peckham, 1991).

The new NHS R&D Programme aimed to provide a coherent framework for various parts of existing expenditure from the DH/NHS R&D budget. In describing the aims of the new programme, Peckham highlighted the importance of meeting the needs of the health service (Peckham, 1991). The new structure was intended to allow the DH/NHS to act as an 'informed customer' by enabling the R&D needs of the NHS to be identified. To this end the new system had wider responsibilities than its predecessors (DH 1991 and 1993; Peckham, 1991; Black, 1997; Harrison and New, 2002). The key issue of how the R&D Programme could perform the role of proxy customer was therefore being addressed. This was particularly important given that at much the same time, and again driven by the new thinking about the organisation of public services, the NHS achieved a greater degree of autonomy from the DH.

Supporting the post of Director of R&D, a new Central Research and Development Committee (CRDC) was also established in 1991. It was composed of 17 doctors, two nurses, three NHS managers, two social scientists, a medical physicist, an industrialist and one lay person. Two thirds of the members were academics (Black, 1997). One task for the CRDC was to advise on priorities for R&D, taking account of the views and interests of a range of professional and patient groups, the government departments responsible for health, and NHS commissioners and providers. Once the CRDC had identified a topic to be a national priority meriting time-limited funding from a central budget, an Advisory Group would recommend the specific areas that would benefit from research, consulting widely with the NHS. Then a Commissioning Group would invite bids and make funding recommendations. The first Advisory Group was for mental health and learning disability, the last for methods to implement R&D findings. These two illustrate the varying nature of the topics - which ranged from specific disease areas to more generic organisational issues (DH, 1993). The Advisory and Commissioning Groups consisted of a mixture of researchers and customers in the relevant field, but it was not always easy to coordinate the two perspectives. Neither was there always agreement about the appropriate methodologies to use (Hanney et al., 2003a). These groups were perhaps the central element of the new R&D strategy and, as reported by Harrison and New (2002), when the House of Lords re-examined the issues they broadly welcomed the progress made (House of Lords, 1995).

The department's centrally managed programme continued under a new name: the Policy Research Programme (PRP). Some of the elements associated with it, such as the research units and certain RLGs, were retained, even though the House of Lords reported that the latter 'seldom meet' (House of Lords, 1988, para 1.24). The RLGs had varying success: some were strongly criticised, while others, including that for children's services (DH, 1994), were seen as allowing researchers, research managers and the policy divisions to work in a productively collaborative way.

The PRP now more explicitly served the needs of policy-makers in the DH. The pattern of unit funding inevitably changed over time in relation to departmental priorities and the movement of key staff (DH, 1995). The Williams Review recommended strengthening the long-term research infrastructure by the creation of a small number of larger centres on ten-year contracts (DH, 1992). Whilst the review confirmed that 'Peer reviews are concerned with the quality of research in the units not with its relevance' (DH, 1992, para 44), it also indicated that government funders were changing the emphasis of evaluation. A National Centre for Research and Development in Primary Health Care was set up in Manchester University. Illustrating a continuation of concerns, discussed in Part II, about the balance between quality and relevance, the statement of requirements for the new centre stated that the evaluation of the centre would, 'be based on the principles of peer review of the quality of the research, its responsiveness to customers' needs and the effect of the research in influencing policy decisions and service delivery.' (DH, 1993a, para 9.2).

The need to find effective ways to transfer research findings to the customers in the policy-making bodies underlined the importance of brokerage. This still had an important part to play, especially for the PRP, but to transfer findings more widely to the customers in the NHS, a wide information systems strategy was adopted. It became concerned with issues of knowledge management and this strategy entailed funding the establishment of the Cochrane Centre and the Centre for Reviews and Dissemination (Black, 1997; Peckham, 1999). Subsequently the international Cochrane Collaboration, and then the Campbell Collaboration, were established. The evidence produced through the advanced techniques now used in systematic reviews, or meta-analysis, is generally thought to provide a much stronger basis for informing policy and practice (Macintyre et al., 2001), though others have reservations.

A major feature of the new NHS R&D Strategy was the further integration of R&D into the regional structure of the NHS. The largest element of the NHS R&D funding was money provided to teaching hospitals for 'own account' research and, in particular, the service support costs the NHS incurred when staff conducted research funded by the MRC, Wellcome Trust and other bodies. Such funding became one arm of the regional R&D structure, which also allocated funds in a responsive mode. This structure played a crucial part in attempting to develop the role of the NHS as an 'informed customer' for R&D. In a range of activities, including priority setting, utilisation of research and public engagement, the regional dimension facilitated attempts to build the R&D process more firmly into the customer organisation. Changes in the structure of the NHS itself, to reflect New Public Management approaches such as the purchaser/provider split, meant that the NHS R&D funding streams were reorganised to protect the major research hospitals from the potential danger of becoming seen as expensive health care providers because of research costs (Culyer, 1994). Furthermore, when the regional structure of the NHS itself was amended, and later disbanded, reorganisation of the R&D structure was necessary.

Early 21st century

Recent reviews of aspects of the NHS R&D Programme have made various criticisms (Harrison and New, 2002; Dash, 2003). Harrison and New suggest that although more voices now have a chance to make themselves heard as to the appropriate research priorities, the pattern of expenditure still embodies significant biases of the kind identified by the House of Lords in 1988. It is claimed that this is in part due to a

> failure to take stock of the whole field in which the Department of Health has a role... Lacking a general set of principles to guide the allocation of resources, the Department of Health cannot make systematic decisions about what it funds, or fully justify the decisions it does make. (Harrison and New, 2002, p.111).

This criticism partly builds on a detailed analysis of the difficulty that the DH has had in developing mechanisms to identify what was happening to the large amount of NHS funding, known as Support for Science, used to support the research funded from other sources such as the MRC: 'after the best part of a decade of financial reform, the Department of Health still does not have the financial machinery it requires if it is to influence, in the light of its priorities, the content of all the work it currently funds.' (Harrison and New, 2002, p.77).

They suggest that the reforms of the first few years of the 21st century may lead to improvements. Indeed, it does now seem that after many years of grappling with these issues a system is emerging which enables audits to collect data about the amount of NHS R&D resources spent on different priority areas, including those, such as cardiovascular disease, for which National Service Frameworks have been developed by the NHS (DH, 2004).

A review prepared for the Health Foundation and the Nuffield Trust (Dash, 2003) recently claimed that health services research is often inaccessible to policy makers, clinicians and health care managers. An accompanying consultative document suggests one of the principal challenges facing HSR, particularly in the light of there being multiple funders, is that 'The absence of traditional client-contractor relationships means that the line between funders and beneficiaries will always be blurred.' (Health Foundation, 2003, p.1). It also asserts both the need for 'new roles' - people who could act as brokers or translators - and the importance of 'establishing a client centred approach to commissioning research' (p.10).

As commentators again turn their attention to such issues, the concepts of brokerage and of customer/contractor relationships developed earlier in this book, and its 1983 edition, could be applied to the current situation and the recent analyses of it. As we have noted, a central issue is the role of the customer or client. The prime focus of the Rothschild reforms was to enable the DHSS to better address problems in fulfilling the customer role for research to meet its own policy-making needs. We have shown that achieving this was difficult, and that attempts by the DHSS to act as a proxy customer for the whole NHS and social services field (as well as social security) faced even more difficulties. The emphasis on attempting to incorporate the NHS funding for research into an overall structure marks one major difference between the reforms introduced from 1991 and the Rothschild period. As noted above, Harrison and New (2002) usefully adopted a systemic approach which has enabled them to document in detail the difficulties involved in creating mechanisms to cover the entire system. But this is an area in which progress has now been made, and in which Harrison and New are correct to observe that the government is the only player that can consciously adopt the role of 'system orchestrator' (2002, p.167). Much of the criticism by Dash (2003), and focus for consultation (Health Foundation, 2003), relate to the need for a more collaborative approach towards HSR. The importance of this concept is further addressed in our overall generalisations and conclusions.

There are, however, further related concepts from the 1983 edition that can be usefully applied to the current situation. First, there have been determined efforts to work with the research councils through a series of concordats and there is increased recognition of the value of networks. The UK Clinical Research Collaboration (UKCRC) was established in 2004 to form a network of the major stakeholders that influence clinical research in the UK. Chaired by the Department of Health's Director of R&D, it includes representatives from the main research funding bodies such as the MRC and medical charities, industry, the NHS, academic medicine, other government departments and patients (UKCRC, 2004). It was followed by the Health Services Research Network created in 2005 with the aim of bridging the divide between users, funders and researchers (The NHS Confederation, 2005). Furthermore, much of the growing importance of the EU funding of health research through the series of framework programmes takes the form of research networks.

Second, and returning specifically to DH/NHS R&D, there have been developments that should help counter some of the factors that complicate the customer role. These complicating factors include the multimodality of both government and science and the consequent epistemological issues that arise; they are being addressed by an increased specification of the role and structure of different parts of the R&D Programme. Overall, in addition to the continuing Policy Research Programme, the current NHS R&D Programme consists of two main streams: Support for Science, which has already been described, and Priority and Needs funding. The latter consists of funding that goes to both the NHS institutions directly (and these arrangements are covered by the developing audit arrangements described above), and the National NHS R&D Programme. This national programme, in turn, consists of three main programmes and three cross-cutting ones (DH, 2004a). The contribution of the National Programme, in particular, can benefit from being analysed in terms such as multimodality: the three main programmes relate to different customer needs and utilise different, but overlapping, research disciplines. The three programmes are: Health Technology Assessment (HTA); Service Delivery and Organisation (SDO); and New and Emerging Applications of Technology (NEAT).

The National Coordinating Centre for Health Technology Assessment has responsibility for ensuring there is a wide range of consultation, including with the public, before priorities are set for the HTA Programme. Extensive use is also made of systematic reviews to identify what is already known, whether further primary research is required and, if so, on what aspects it should focus. The HTA Programme funds over £10 million of research each year and

> provides crucial evidence to help the NHS deliver the best possible health care...and its work is used by national standards setting bodies such as the National Institute for Clinical Excellence (NICE), the National Screening Committee and the Modernisation Agency. Its reports feed into guidelines produced by Royal Colleges and specialist societies. (DH, 2004b)

Why has this part of the programme appeared to be successful? First, the Programme has built up a 'reservoir of research' in various fields, as some of the RLGs had hoped to do (see Chapter 5). Second, HTAs are a type of research potentially more amenable than many to being used in policy-making (Hanney et al., 2003), partly because of the link with systematic reviews. Third, the National Institute for Health and Clinical Excellence (NICE) has eventually provided an institutional arrangement to appraise the research evidence, including the many HTA reports that have been produced (DH, 2004), and to make 'rational' policy recommendations. Furthermore, NICE now also provides a 'receptor' body, as described in the 1983 edition (see also Kogan and Henkel, 2000): it specifically acts as a customer and routinely commissions HTAs when it is faced with making specific recommendations in an area where it identifies the need for further research. NICE's methods have themselves recently been favourably reviewed (WHO Europe, 2003).

By themselves HTAs will not necessarily be much used. It is only when the policy-making, or receptor, body is properly established to use the research that it can make a difference (Hanney et al., 2003). A key feature of this is that one receptor body, NICE, specifically gives great authority to some HTAs, thus enhancing the status of the type of knowledge production involved in Health Technology Assessments.

Perhaps partly as a reflection of the feeling that the HTA Programme was proving to be successful, the Service Delivery and Organisation National R&D Programme was formally established in 2000 with a national coordinating centre similar to that of the HTA Programme. There has again been an attempt to identify the research methods that will be most appropriate for the specific research areas (Fulop et al., 2003) and to explore how the more diverse forms of knowledge produced in this field can be synthesised (Mays et al., 2001). There is, however, much controversy about some of these issues, and a major task facing the SDO Programme is developing links with receptor bodies that will recognise the research produced as having an equivalent authority for its purposes as does that produced by the HTA Programme for its purposes.

The main purpose of the NEAT Programme is to fill a perceived gap between the other research streams by addressing the development barrier. It does this by supporting work which applies recent advances in fundamental knowledge and technology to the development of new products and interventions for improved health and social care or for disease prevention and treatment.

The three cross-cutting programmes within the National NHS R&D Programme are: Methodology; Research Capacity Development; and INVOLVE – promoting public involvement in NHS, public health and social care research (DH, 2004a). The first two reflect a concern to address the longstanding issues of developing authoritative research methods appropriate to their tasks and enhancing the relevant research capacity. The third illustrates the growing concern to meet the demand for public involvement. It was established to

> promote public involvement in research, in order to improve the way that research is prioritised, commissioned, undertaken, communicated and used. We believe that the active involvement of the public in the research process leads to research that is more relevant to people and is more likely to be used. (INVOLVE, 2005)

Such developments not only represent part of the more general trend, identified in Chapter 3, toward greater 'customer responsiveness' in public services, but also indicate a broadening of the definition of the term customer and inevitably involve introducing additional forms of knowledge into debates about research. Even within the MRC, a body where the internalist approach to research agenda setting is still strong, the Consumers' Liaison Group was established with a broad brief to supply lay perspectives. Whilst achieving apparently only minor changes in procedures and documents, crucially it is playing a part 'in a perceived re-orientation of organisational culture towards greater accountability and transparency' (Milewa et al., forthcoming). The precise nature of public involvement in NHS research not only takes various forms but faces many barriers with, as yet, 'little evidence of consumer involvement influencing research agendas' (Oliver et al., 2004).

Closely linked to the push to take more notice of customers, one driving force running throughout recurrent phases of reform since 1991 has been the need for the research to be translated into practice: 'the NHS...should ensure that the fruits of research are systematically transferred into service' (House of Lords, 1988, para 4.1). In describing the weaknesses that the new system hoped to address, Peckham opined that, 'Too little attention has been paid to the dissemination of information in such a way as to maximise the chance of securing uptake.' (Peckham, 1991, p.370). It has been argued that it is the attempt to develop mechanisms and networks to facilitate the greater use of health research that distinguishes a health research *system* from earlier approaches which relied so much on the traditional individualistic determination of medical research priorities in universities and hospitals (Hanney et al., 2003).

In this context it was seen as important for the R&D Programme to start commissioning evaluative work into the benefits or payback from health R&D (Peckham, 1993). This work indicated that the system was producing research that was having benefits across much of the range of payback categories, though problems were also identified (Buxton and Hanney, 1996; Buxton et al., 2000). The stream of work on R&D payback was further developed to show how a regular monitoring system could be constructed and incorporate the different kinds of benefits that can be expected from research (Croxson et al., 2001). Furthermore, underlining the shift towards assessment that strives for more balance between scientific quality and relevance, the payback approach is now referred to in guidance given to Directors of DH-funded units and programmes about preparing for their DH review. Nevertheless, any consideration of the role of assessment of government funded research must accept that the pressures created in the 1990s by the Research Assessment Exercise (Henkel, 2000) have in many ways reinforced the traditional emphasis on the academic quality of the work as discussed in Chapters 9 and 10. Such pressures are keenly felt in major centres of health research (Hanney et al., 2000).

Overall, therefore, there are now pressures to extend the notion of the customer and broaden forms of research assessment at the same time as some of the enduring criticisms about the difficulties of producing research relevant to the needs of users are being repeated. Despite the difficulties, the customer voice has been strengthened and technical resources for conducting research and developing evidence useful for policy and practice are being been created. Many issues remain unresolved in an ever evolving system.

Generalisations and Conclusions

What generalisations can we draw from this institutional history of the Rothschild period and its sequel? In this depiction we employed a range of formulations and can now consider how far they need to be refined or revised.

We have depicted our story as being of two complex institutions, endeavouring to collaborate, and encountering difficulties in doing so. We can follow two tracks: the technical issues concerned, the complexity and plurality of modes and the resulting structures and roles; and the socio-political dimensions as evinced in the relationships between power, knowledge and values. In analysing these themes, we can ask how far recent history causes us to change or extend the earlier analysis.

Multimodality of government and science

Our descriptions and analyses of the system as it developed from the 1980s continue to challenge the simple models of authoritative and single-minded government and self-regulating, internally determined science which we depicted in Chapter 1. On the multimodality of government and science, we note that government and science contain elements of convergence and divergence. The boundaries of both are permeable and moving. The enormous complexities of each, both within themselves and in relationship with each other, have demanded a shift in emphasis from internal expertise to boundary crossing and brokerage.

Science is variable in its relationship with those in government who would 'steer it'. There are degrees of 'finalisation' and different starting-points: some from the rigours of academic disciplines, others from social problems and domains of concern.

Because government and science are both multimodal and change as they interact with each other, it becomes clear that no single or simple model of science policy is appropriate, and much of what we have observed throws doubt upon some of the doctrines enunciated in the Rothschild Report as well as their radical modification from the 1980s onwards. Multimodality, permeability and moving boundaries are thus the institutional characteristics of both government and science.

Policies change rapidly as do their environments. In order for government to present objectives for joint work with scientists it has had to have clear policy directions within its own discrete policy divisions; each of these have been subject to a host of political, client group, and economic pressures that change rapidly even over such short periods as those of our original study.

Since the first edition, as noted above, there have been further shifts in the balance between science and government with an increasing number of policy-making levels, within as well as beyond national boundaries, and a wider range of actors accompanied by a greater variety of formal and informal modes of participation. The multi-level, multi-actor approaches serve to focus attention on this growing complexity in which the multimodality of science and of government is intensified. Widening participation not only makes the arena more populated, but also brings in different forms of knowledge, in particular the perspectives of the general public. Both government and science are faced with these pressures and some of the attempts to engage the public with science involve them being consulted about policy-making in controversial areas involving science, such as human reproduction. However, despite the growing importance of 'hybrid communities' for the production of knowledge, science retains considerable power over how problems are defined and therefore within what scientific or cognitive frameworks they should be addressed (Weingart, 1997).

Whilst there are indications that to some extent the issue of multimodality is now being addressed, and it is more clearly part of government's normative furniture, there are now even more facets to it than applied at the time of the first edition. This has implications for how the role of the customer for research can be conceptualised and for the question of which groups have power over the research agenda.

Customers and contractors

The role of the customer has become even more salient but it had always entailed intelligence gathering, commissioning and evaluation of research, assimilating results as a receptor, assisting knowledge transfer, dissemination. A more recent elaboration has been the involvement of users of research, either as practitioners or patients and other beneficiaries. Working on behalf of users is the more complicated by the duality of the direct and the proxy customer.

Most important have been the attempts to recognise the NHS, in addition to the Department itself, as a customer. The metaphor of the customer had assumed that government could be clear and authoritative in stating its objectives before entering the marketplace to purchase knowledge from researchers. In the Rothschild period, the department itself modified the metaphor and tried hard to enter into more reflexive working relationships with the scientists. The 1970s story of the Panel on Medical Research, the Small Grants Committee, the encounters with science in the RLGs, all exemplified a government department seeking accommodations with science. In particular, the Department's relationships with the MRC changed course sharply three times in less than nine years as the customers first advanced and then retreated under pressure from the medical scientists. But in the 1970s the constraints on the Department offering a decisive and unambiguous framework for scientific development were too strong, and its capacity to formulate policies that can provide a full frame for research remains uncertain.

The metaphor of the contractor, too, assumed a coherent and independent scientific community which, being in command of its own values, would be capable of imposing demonstrable patterns of intellectual authority upon those who were part of it, and those who would use its products. It would also have command over its own economy.

The original customer-contractor relationship assumed negotiation and exchange between two sets of institutions. In this exchange, the purposes of commissioned research would be stated by policy-makers, but with two important reservations. First, the purposes would be worked out collaboratively with scientists and, secondly, the purposes laid down by the customers would not be the only objectives of science. They would include those emanating from the demands of the disciplines themselves.

But the position of researchers has changed. At the time that the 1971 Rothschild Report was written higher education was still to expand and researchers would either be on tenure, or have a promise of tenure, or of relatively fluent access to alternative sources of funds. The evaluation pressures on research units have been increasing since the 1970s and, more recently, the demands of the Research Assessment Exercise, based on 'scientific' and disciplinary criteria, are ever more strongly asserted (Henkel, 2000) while the R&D Programme insists on the attention to be given to the benefits from research and it is claimed that traditional academic perspectives are too dominant (Dash, 2003).

A specific element of the customer role concerns brokerage, which we depicted as research management, acting as liaison officers between research and the policy divisions. Brokerage and its manifestations are still at work and still have an impact on the take-up of research (Henkel, 1994), and, as noted, it is being increasingly strongly promoted as a key element in the success of a research system.

Two new factors have emerged on the scene since the 1980s. First, some mega programmes - perhaps the human genome programme is the obvious example – are so important and large, and international in their reference groups, that it is unlikely that government could exercise a command position over them once funding is allocated. The Department nevertheless established a mechanism for gathering intelligence on the human genome programme's implications for health policy. Secondly, scientists collaborating with industry, as with the Foresight Initiative, have a second source of legitimacy beyond their government funders and although Foresight had little impact initially on health and life sciences (Henkel et al., 2000; Hanney et al., 2001) it is argued that the industrial perspective is of increasing importance in the NHS R&D programme (McNally et al., 2003).

Government- scientist relationships at the macro-level

Underlying the nature of the systems created and recreated during our period is the question of government-scientific relationships at the macro level. The customer-contractor relationship implied exchange, but the relationship between customers and contractors was distorted by imbalances of power between government and science and within science itself. In the Rothschild system, many of the 90 or so scientific advisers began with a strong commitment to make the 1970s system work. They hoped to do something to improve the research programme, to increase public support for under-regarded areas of science, and to raise standards of research supported by the DHSS. Government would seek their advice and policy-makers would give encouragement for neglected work that needed to be done.

In return for whatever benefits it conferred on science, government certainly received good advice within the RLGs and in individual projects, if not at the macro level. Part of the exchange was also the legitimation of the government's commissioning policies and decisions. Even if the machinery never fully worked, government could feel that work processed by RLGs, within the rules of the CSRC, would be scientifically sound as well as socially useful. Government would be less vulnerable to attack as closed and impervious.

Imbalance in exchange relationships generates power (Blau, 1964) and the exchanges did not all favour one side. Basic scientists could withhold approval of what government was doing; their disapproval eventually affected the shape of the science-government encounter. They eventually loosened the DHSS's grip over policy-related biomedical research, through the transferred funds mechanism. Government's Concordat with the MRC (1980) was a victory for the medical scientists. Their disapproval had weakened the DHSS's power quite early. As we have noted, the DH's capacity to direct the policy frame remains in doubt.

Issues of knowledge and power

In recent years, the scientific position has both weakened and strengthened. It has been weakened by government control over scientists' academic bases and perhaps the increasing number of perspectives informing the research agenda, but also strengthened by the creation of new technical resources contributing to policy and practice, as with some HTAs and the NICE. The attempts to strengthen the evidential base through systematic reviews and the work of the Cochrane Collaboration potentially enhance the role of science, and these factors are seen internationally as being of considerable importance for the improvement of health systems (Lavis et al., 2004).

Our original research displayed characteristics of the different scientific groups which we believe still persist. The varying degrees of power and independence possessed by different kinds of scientists can be explicated within three interconnecting parameters: the institutional and those of value and of knowledge. The MRC came through the experience of the Rothschild experiment stronger than before. Its internalist view of the authority of science was accepted and the DHSS was compelled to continue to negotiate for its concerns with virtually independent medical scientists. The contrast with DHSS-funded researchers or the (as was) Social Science Research Council was striking. For the most part, they tried to advance knowledge in areas where the authority of science was less secure without the resources and the institutional status enjoyed by the MRC.

As to the differences in social values, the 'DHSS' scientists assumed that knowledge should contribute to policy-making, that policy-makers should listen to scientists and collaborate with them and that they, reciprocally, would share the quest for objectives with policy-makers. By contrast, the 'MRC' scientists, while committed to research that would eventually improve the human condition, believed that they benefited mankind most when they listened to the messages coming from their own clinical experiences and from scientific development.

The knowledge which the DHSS sought to sponsor can be viewed along two dimensions. First, its reference groups and its areas of interest were broader and more diffuse than those of the science sponsored by the MRC. Whilst much DHSS science remained within academic disciplinary frameworks the reference groups consisted not only of the academic-peer system but also of policy-makers, practitioners and client groups to whom their work is ultimately directed. And some researchers moved away from the firm disciplinary bases and into domains of interest where the starting-point is a social problem, to be tackled by reference to a wide range of disciplinary perspectives, but ultimately responding to networks, connections and logics not easily defined in terms of academic disciplines.

The second dimension of knowledge concerns the stage of 'finalisation' reached. We share the intuition of Weingart and others (van den Daele et al., 1977; Weingart, 1997) that the susceptibility of science to 'steerage' by policy sponsors depends upon the stage of development reached by the science.

There were successes in the science-government encounter. The RLGs did identify policy-makers' research needs. They found researchers, maintained research standards, and some concerned themselves with dissemination and implications of research findings for policy. But we have shown that the limits of RLGs were in part responsible for their success. Policy-makers did not want too much elaboration of policy problems. They did not encourage the crossing of boundaries nor the examination of the more fundamental causes of problems. The function of government is to reduce and not elaborate problems, and the policy-makers kept to this remit.

Reinforcement and improvement of relationships between the MRC and DH tend to reinforce arguments about the relationships between epistemic communities and power. The MRC has been able to make claims of its location in both 'public health' and 'HSR', whereas it remains difficult for the Economic and Social Research Council to gain a strong position. This difference may be conceptualised in terms of restricted and unrestricted science – see Chapter 2.

The case for adopting interactive models

If different sciences and different modes of scientific work have different impacts upon government those thinking about the organisation and exploitation of science by government have to make a sensitive adjustment which takes account of the levels of application of the knowledge sought, the models of policy creation and implementation assumed, and the kind of science appropriate for the particular problems being tackled. Against such variables, it is then possible to begin to model the nature of the relationships that will work best, the forms of institutionalisation to be encouraged or endowed and the forms of mediation and brokerage required so that boundaries can be crossed or moved.

Several attempts have been made to model, for both normative and analytic purposes, the relationships between government and science (Caplan et al., 1975; Weiss, 1977; Davies et al., 2000; Denis and Lomas, 2003). Those concerned to maximise the linkage adopt consensual and linear models and tend to emphasise management or steerage. Models, however, which assume that knowledge percolates rather than prescribes advance interaction, multidimensionality and sensitive reference to contexts. Some emphasise that research helps policy to discard or falsify unproductive patterns of intervention.

We come to a general if biased conclusion. Policies can only be implemented successfully and practice improved in the areas of health and social welfare if those who seek to analyse issues and suggest solutions are interactive and reflexive. Government must be authoritative as it determines its policy priorities and allocates money to researchers. But the concepts to which it can better apply its energies are interaction rather than steerage; impact, implementation, policy communities and domains of concern, rather than rationalistic or imperative planning.

One reason for this is that the commissioning of research is subject to its own natural history. At the beginning, it might be concerned with producing much needed but relatively simple data which sketch social and other conditions. But as policies begin to work through the system and their impacts and effectiveness become the focus of interest, so the research becomes more multivariant, and the issues involved become more complicated in terms of their intellectual content and their political connotations. Research concerned with the 'outputs' rather than the 'inputs' end of the policy process (Ruin, 1984) is, therefore, likely to add more to divergence, pluralism and just plain confusion than to clear-minded and authoritative closure of the policy process. As the problems become more complex, so does science and so does the task of government in promoting and using policy-relevant research. If it clears up its own complexities, a government department will then have the problem of how to act within larger networks.

A second reason for adopting interactive models is that much of the knowledge required by government is the kind of knowledge that pressure groups and political opponents as well as field authorities and practitioners might use whether or not it is commissioned by government. Knowledge affects the consciousness of many groups in society acting through a process of enlightenment and social conflict or serving as ammunition for those engaged in political and social conflict (Weiss, 1977). Indeed, wise government provides succour for those who test its actions and provide its counter-analysis.

The collaboration model has been, in fact, an important mode in government's commissioning of science. The characteristic pattern is that of the negotiated contract. In such a relationship, the patterns of learning are interactive. The scientist does not rely upon the internalised norms of his profession to determine the objectives of research but works with the policy-maker and practitioner to evaluate zones of concern and to find the appropriate methods of inquiry. For government, the collaboration may be formative in that policy issues become clarified as they are stated as problems. At the same time, however, the closer the collaboration between policy-makers and researchers, the more will problems become elaborated and complexity rather than reduction and simplicity established. This was the dominant mode that the DHSS adopted even if it shrank from its full implications. We have seen that it led to the elaboration of new kinds of customer behaviour and of new forms of brokerage roles. The brokerage roles can be structural, as through working parties or committees, or they can reside in individual roles such as those of liaison officers or, most conspicuously, the Chief Scientist. Many of the recent demands for reforms are, whatever the actors are currently called, to go further in these very directions.

Conclusions

It will be appropriate to close this book by reflecting on the justification of the pursuit of knowledge as an aid to policy. Scientific development must rest not only on the need for greater technical knowledge but, we have asserted, on the need for policy-makers to establish the range of uncertainty, and of the unevenness of development which inhabit their domains. Science, broadly and properly construed, helps establish the uncertainties and ways of reducing them into logical and hence manageable proportions.

Long experience now of the customer role suggests in some circumstances the system is working, and in other circumstances not working so well.

There is clear support in the UK and internationally for the collaborative approach that was discussed in the first edition (see Denis and Lomas, 2003; Walter et al., 2003; NAO, 2003; Haines et al., 2004). Greater efforts are required to develop collaborative approaches if the full potential of health research to influence policy is to be realised. However, as we have seen, the impact of the HTA research on NICE policy appraisals is not necessarily dependent upon interaction between the researchers and policy-makers, at least after the priorities have been established. This points to the importance of multimodal analysis when considering the situation in the UK.

The analyses in the two editions of this book suggest that a systems approach should be adopted towards health research within a country, but that a disaggregated multimodal approach is then required for different elements within the system. The partnership or collaborative approach is seen to be a good way to engage researchers and policy-makers in identifying priorities and developing research ideas. Although large-scale priority setting mechanisms have a role, the direct involvement of those responsible for specific policies is the most likely way to increase the chance of producing research that will have an impact.

The incentives for researchers to engage in activities linked to achieving research impact also have to be considered, as do the organisation of the receptor body and the need to use appropriate research methods for the different tasks. Experience from the 1970s shows that it requires considerable commitment from policy-makers if they are to play the role of 'informed customer', and even then it might be possible for this to happen over only a part of the policy agenda. Furthermore, it is likely to involve increased resources being devoted to research management and brokerage roles that are always vulnerable to criticism about excessive expenditure on 'unproductive' bureaucracy.

Finally, we must remind the reader that much of our material here derives from sustained empirical research completed at the end of the 1970s. We have not been able to continue the first hand research into the later period but have instead relied mainly on secondary sources. Yet, it remains clear that constructs and lessons learned in the earlier research remain relevant today. Thus themes we were able to follow in the late 1970s remain fruitful areas for further external disciplined enquiry: the interplay between government and its scientific advisers; the current role and state of brokerage; and the story of what happens to research once received by the customers. Particularly in the context of the growing international focus on these issues, as identified in the Introduction to this second edition, it is important that policy developments in the field of the organisation of research are themselves informed by detailed research and analysis.

Appendix: Preface to the First Edition

THIS BOOK is about government's attempt to commission science. Its starting point was that in 1974 we were invited by the Research Management Division (now called the Office of the Chief Scientist) of the Department of Health and Social Security (DHSS) to study the research management system which had been installed following the Rothschild Report.

Whilst the invitation was to advise the Department, from the beginning it was agreed that we should eventually move from the role of consultants into the mode of independent researchers making our findings available to the wider scholarly public. We have had every assistance from the DHSS in achieving that end.

One of us, Maurice Kogan, worked on the project part-time throughout, first with Nancy Korman who was, for five years, the full-time researcher for the first part of this study. In the latter part of the study Mary Henkel replaced Nancy Korman when she took up research elsewhere. Many of the judgements in the book, and much of the empirical work, particularly in Part II, are based on field work, analysis and writing by Nancy Korman and we are greatly in her debt for her direct and indirect contributions to our account.

Our methods have responded eclectically to both the requests made by our DHSS 'customers' and the momentum which the research developed of its own accord. In the first period of our study (from 1974 to 1979) a principal product was a series of papers evaluating the working of the research commissioning system. This involved attendance at virtually all of the meetings held under the aegis of the Chief Scientist, that is to say, meetings of the Chief Scientist's Research Committee (CSRC), its three intermediate boards concerned with medical, personal social services and health service research, the Small Grants Committee and Research Liaison Groups. All committee papers were made available to us and they, together with the minutes of meetings, are essential records of the events which we now analyse. This documentation was supplemented in many cases by brief evaluative notes of the most important events and by interviews conducted with some of the main actors.

Many of these were cleared with interviewees, but more systematic verification took place when the series of evaluative papers was submitted for comment to research management and, where the issues warranted it, to the committees upon which scientific advisers sat. The data from the first part of the study were collated and analysed in our first publication, *Government's Commissioning of Research* (Kogan et al., 1980).

A second phase began in 1980 when we became privileged observers of the Chief Scientist's visitations to five DHSS-funded research units. We interviewed members of the units and of the DHSS staff concerned with them both before and after the visits and attended both the formal parts of the visitations and the Chief Scientist's visiting experts' private meetings. We also engaged in 'light contact' with five further units not being visited. The invitation from the Chief Scientist, Professor Arthur Buller, to undertake this review and his encouragement to publish our findings, although largely inimicable to the view of science and its relationship with policy advanced by him, reflect great credit on him and on the DHSS.

On the basis of this second period of research, Mary Henkel and Maurice Kogan published *The DHSS Funded Research Units: the Process of Review* (1981). But events did not stand still and while this specialist study was being undertaken drastic changes were visited on the research commissioning system. It was therefore necessary to decide an arbitrary end-point for our history of the Rothschild enterprise within the DHSS, and we called a halt to our history in April 1981. We conducted some interviews after that date and held some further meetings in the Department to verify our findings from the study of the units and also to verify some of the data contained in this book.

Between 1974 and 1981 the three researchers conducted a total of 208 interviews with policy-makers and research managers in the DHSS, scientific advisers and researchers, and representatives of health and local authorities. These interviews were mainly, but not always, with individuals alone. During the first period of our research we attended a total of 210 departmental meetings as non-participant observers. These included eleven meetings of the Chief Scientist's Research Committee, six of the Panel on Medical Research, eight of the Intermediate Boards, some 156 of the Research Liaison Groups (almost all the meetings held between 1974 and 1979 of eight of these eleven groups), 21 of the Small Grants Committee, and eight of Research Management. In the second period we observed five internal departmental meetings connected with Chief Scientist's visits to funded research units, in addition to the visits themselves. A full breakdown of the interviews conducted for this part of the research can be found in the Appendix to Chapter 8.

We have drawn freely on the papers and minutes from meetings of the whole period 1974 to 1981. We were also given liberal access to some internal departmental papers and references to some of them will be found, in suitably coded forms so as to preserve civil service anonymity rules, in the references at the end of this book.

We could not tackle the whole range of DHSS research, illuminating though comparisons between health and social service and social security research would have been. Nor until we examined the unit review process were we able to broach the issue of customer reception and treatment of research. We did not take our study into comparisons between the DHSS's mode of working and those of the MRC although both of these possibilities were raised early in our study. Fortunately the MRC was eloquently represented at the Public Accounts Committee and its attitudes and *modus operandi* are a matter of public record; one of us was a member of the SSRC and chairman of its Health Studies Panel during part of the period of our study and our brief references to these are based on direct observation. In all, we have been opportunistic in securing access to the field and have not tried to break down doors that were not already open to us. On the whole, we have been privileged to see and hear that which we now record. In one area where we did undertake research, the DHSS's commissioning of R & D in computers, Nancy Korman undertook a great deal of field work and produced internal papers. In the interests, however, of thematic unity and economy that fascinating set of issues is excluded from this account.

Some of our omissions sit more comfortably because of the excellent work done by others. Work published by M. D. Gordon and A. J. Meadows of the Primary Communications Research Centre, University of Leicester, (1981) analysed the effects of research commissioning on the researchers whilst pursuing its main purpose of assessing the dissemination of findings of DHSS-funded research. In 1977 Professor Louis Moss undertook a study of the DHSS's use of its research units as a principal research resource and this produced material, which we exploit, on the research perceived to be necessary by DHSS administrators and professionals. Both of these studies were commissioned by the DHSS at the same time as our own, and we commend them as complementary to our own work. Other academic studies have tackled areas outside our own somewhat esoteric remit. The inner life of research units has been well recounted by Jennifer Platt (1976). Broader issues of the sociology of science have been discussed by Stuart Blume (1974; 1977) and the social organisation of science has been treated in masterly fashion by Michael Mulkay (1977). The Social Science Research Council funded a study of research careers and funding conducted by a team at Goldsmiths' College, University of London, and Patricia Thomas (1982) has examined the relationship between social research and government policy by tracing connections between SSRC and foundation supported projects and the policy-making process. These studies help provide an essential context within which our own work has been conducted. In addition, we benefited from membership of a seminar organised under the aegis of the Goldsmiths' College Project and convened by Matthew Melliar-Smith at which other students of research and science policy presented papers and offered critique of current work, including our own.

We make one bold claim for this study. Our account builds on the literature of the science of science and public administration. The science of science contains important studies of the epistemology, sociology and social organisation of science, But the reader of such key journals as *Minerva or Knowledge* will find it hard to discover empirical studies, as opposed to normative models, of the way in which science is perceived, steered or influenced by public policy-makers or of the relationships between policy systems and the science which they finance. We hope that this account of our work within the DHSS and its associated research enterprises provides such a study and also contributes to developing theory in an exciting field.

In a study that has taken over seven years to research and write we have engendered many debts. We acknowledge the patient help received from many members of the DHSS staff, from scientists and others involved in the Chief Scientist's system, and from researchers who were willing to be interviewed and to contribute in other ways to our study. A particularly heavy burden was carried by friends and colleagues who critically and helpfully read this book in draft and our earlier papers, *Government's Commissioning of Research: A Case Study* (Maurice Kogan, Nancy Korman, and Mary Henkel, 1980) and *The DHSS Funded Research Units: The Process of Review* (Mary Henkel and Maurice Kogan, 1981). The many people within the DHSS who helped in this way cannot be named because of the civil service convention of anonymity. While they did not expect us to conform to departmental views of what we observed, this rule meant that they could be helpful to us without committing the DHSS to anything that we were likely to say in our published reports. We received helpful advice from Professor Tony Becher, Dr Helen Bolderson, Patricia Broadfoot, Valerie Heyes, Dr Donald Irvine, Daphne Johnson, Tim Packwood, Ellie Scrivens and David Shapiro and, more recently, from Martin Buxton, Eskil Bjorklund, Warren Kinston, Rune Premfors, Stewart Ranson, Olof Ruin, Brian Salter and Bjorn Wittrock. The project and the book were seen through by the expert administrative and secretarial work put into it by Sally Harris.

Bibliography

Key to DHSS committee papers and to researchers' unpublished papers and notes of interviews and meetings

Uncirculated papers and interview notes

BRMP ( )/	Brunel research management project, project papers
FN ( )/	Field notes
ID ( )/	Interviews with department officers
IE ( )/	Interviews with external advisers
IM ( )/	DHSS internal memoranda We compiled no complete archive of all such memoranda but those to which we have referred are numbered serially in each of their years.
IR ( )/	Interviews with researchers

In addition, there were minuted internal research management and OCS meetings as follows:

DHSS MG ( )/	Management group
RP	Research policy meeting
RP (M)	Research policy meeting Minutes

Committee papers

CSRC ( )/	Papers presented to the Chief Scientist's research committee denoting year and series numbers.
CSRC (M)	CSRC minutes (some are not numbered serially in each year and dates of meetings are given in those cases).
CSRC (HPSS) ( )/	Health and personal social services committee papers.
MRC ( )/	Medical research council papers.
PMR ( )/	Panel on medical research papers.
PMR (M) ( )/	Panel on medical research minutes.
RLG ( )/	Research liaison group papers.
RLG (M)	Research liaison group minutes.

The identity of RLGs is denoted by:

CH	Children
SHB	Elderly
MI	Mental illness
MH	Mental handicap
HA	Homelessness and addictions
SGC	Small Grants Committee papers

Official publications

DHSS publications

Better Services for the Mentally Handicapped 1971, White Paper, Cmnd 4683, HMSO, London.

Better Services for the Mentally Ill 1975, White Paper, Cmnd 6233, HMSO, London.

Prevention and Health 1977, White Paper, Cmnd 7047, HMSO, London.

Handbooks of Research and Development 1974 to 1981, HMSO, London.

DHSS Priorities for Health and Social Services 1976, consultative document, HMSO, London.

Three Chairmen's Report 1976. *Regional Chairmen's Inquiry into the working of the DHSS in relation to Regional Health Authorities,* DHSS.

The Way Forward 1977, consultative document, HMSO, London.

DHSS Handbook 1979, *Mental handicap RLG strategy statement.*

Church House Meeting 1979, circulated report of Chief Scientist's meeting with research unit directors, March 1979 (unpublished).

DHSS, SHHD and MRC 1980, Arrangements for the Cooperation of the Health Departments and the Medical Research Council.

The Support of Health and Personal Social Services Research 1982, a report by the Chief Scientist's advisory group,.

DHSS 1988, Memorandum to House of Lords Select Committee on Science and Technology, *Priorities in Medical Research* 1987-8 3rd report, vol II, HMSO, London.

DH publications

DH, 1991, *Research for Health: A Research and Development strategy for the NHS,* HMSO, London.

DH, 1992, *Review of the Role of Department of Health Funded Research Units: The Williams Report,* Department of Health, London.

DH, 1993, *Research for Health*, Department of Health, London.

DH, 1993a, *A Centre in Research and Development in Primary Health Care: A Statement of Requirements*, Department of Health, London.

DH, 1994, *A Wider Strategy for Research and Development Relating to Personal Social Services,* HMSO, London.

DH, 1995, Centrally Commissioned Research Programme, Department of Health, London.

DH, 2004, *Summary Report: NHS R&D annual reporting 2002/03,* Department of Health,London,[http://www.dh.gov.uk/PublicationsAndStatistics/PublicationsResearchAndDevelopment/InformationFromResearch/InformationFromResearchArticle/fs/en?CONTENT_ID=4078384&chk=34ZRnd Accessed 3 July 2004].

DH, 2004a, *National NHS R&D Programme,* Department of Health, London, [http://www.dh.gov.uk/PolicyAndGuidance/ResearchAndDevelopment/NHSLightHouseProject/fs/en Accessed 27 June 2004].

DH, 2004b, *Health Technology Assessment Programme: Annual Report 2003,* Department of Health, London.

Other Official Reports and Documents

Chancellor of the Duchy of Lancaster (OST), 1993, White Paper on Science Policy: *Realising Our Potential: A Strategy for Science, Engineering and Technology,* Cm. 2259, HMSO, London.

Collaborative Training Program, 2004, *Health Research for Policy, Action and Practice. Resource Modules,* The Alliance for Health Policy and Systems Research, The International Clinical Epidemiology Network (INCLEN) Trust, The Council on Health Research for Development, The Global Forum for Health Research. [Available at: http://www.alliance-hpsr.org/jahia/Jahia/pid/38]

Culyer, A.J., 1994, *Supporting Research and Development in the NHS. Report of the Department of Health Research and Development Task Force,* HMSO, London.

Dainton Report 1971, *Report of a Study on the Support of Scientific Research in the Universities,* Cmnd 4798, HMSO, London.

Dainton Report 1971, *The Future of the Research Council System,* Cmnd 4814; *Report of A Council for Science Policy Working Group* 1971, Cmnd 4814, HMSO, London (reprinted 1972).

Framework for Government Research and Development 1972, White Paper, Cmnd 5406, HMSO, London.

A Framework for Government Research and Development 1979, White Paper, Cmnd 7499, HMSO, London.

Fulton Report 1968, *The Civil Service,* Report of a committee under the chairmanship of Lord Fulton, Cmnd 3638, HMSO, London.

Global Forum for Health Research: *The 10/90 Report on Health Research 2001-2002.* Geneva, World Health Organization 2002.

Haldane Report 1918, *Report of the Machinery of Government Committee,* Cmnd 9230, HMSO, London.

House of Commons. *Select Committee on Estimates*, 6th Report, Session 1957-58, Treasury Control of Expenditure, 254, July, para. 23.

House of Commons. Public Accounts Committee 1979, *First Report from the Committee of Public Accounts,* HMS0, London, HC173.

House of Lords Select Committee on Science and Technology, *Science and Government* 1981-2, lst report, vol. 1, HMSO.

House of Lords Select Committee on Science and Technology, 1988, *Priorities in Medical Research* 1987-8 3rd report, vol 1, HMSO, London.

House of Lords Select Committee on Science and Technology, 1995, *Medical Research and the NHS reforms,* HMSO, London.

Joint Health Departments/MRC Working Group Paper, 1973, 'Arrangements for cooperation in the field of biomedical research'.

Medical Research Council, 1988, Memorandum to House of Lords Select Committee on Science and Technology, *Priorities in Medical Research* 1987-8 3rd report, vol II, HMSO, London.

National Audit Office, 2003, *Getting the Evidence: Using Research in Policy Making,* HC 586-1, 2002-3, National Audit Office, London.

Plowden Report 1961, *The Control of Public Expenditure,* Cmnd 1432, HMSO, London.

Rothschild Report 1971, *The Organisation and Management of Government R and D,* Cmnd 4814, HMSO, London.

Rothschild Report 1982, *An Inquiry into the Social Science Research Council,* Cmnd 2554, HMSO, London, May.

Royal Commission on the National Health Service 1978, *The Working of the National Health Service,* Research Report No. 1, HMSO, London.

Skeffington Report 1969, *People and Planning*, report of the committee on public participation in planning, HMSO, London.

The Reorganisation of Central Government 1970, White Paper, Cmnd 4506, HMSO, London.

World Health Organisation, 2004, *Knowledge for Better Health,* WHO, Geneva.

World Health Organization, Europe, 2003, *Technology Appraisal Programme of the National Institute for Clinical Excellence: A Review by WHO,* WHO Regional Office for Europe, Copenhagen.

Zuckerman Report 1961, *Report of the Committee on the Management and Control of Research and Development,* Office of the Minister for Science, HMSO, London.

Other published works

Arnstein, S.R., 1969, 'A Ladder of citizen participation', *Am Inst Planners J,* vol. 35, pp.216-24.

Bachrach, P., Baratz, J.M., 1970, *Power and Poverty: Theory and Practice,* Oxford University Press, Oxford.

Bailey, S.K., Mosher, E.K., ESEA 1968, *The Office of Education Administers a* Law, Syracuse University Press, New York.

Banks, G.T., 1979, 'Programme budgeting in the DHSS', in Booth, T.A., (ed) *Planning for Welfare, Social Policy and the Expenditure Process,* Basil Blackwell and Martin Robertson, Oxford.

Barnes, J., Connelly, N., (eds) 1978, *Social Care Research,* Bedford Square Press, London.

Becher, T., Kogan, M., 1980, *Process and Structure in Higher Education,* Heinemann, London.

Bell, D., 1973, *The Coming of the Post-Industrial Society,* Basic Books, New York.

Ben David, J., 1964, 'Scientific growth: a sociological view', *Minerva,* vol. 2, no. 4, Summer, pp. 455-76.

Ben David, J., 1965, 'The scientific role: the conditions of the establishment in Europe', *Minerva,* vol. 4, no. 1, Autumn, pp. 15-54.

Benveniste, G., 1982, *Professionalisation of Policy Experts,* Tenth World Congress of Sociology, Mexico City.

Berlin, I., 1953, *The Hedgehog and the Fox,* Weidenfeld & Nicolson, London.

Bernal, J.D., 1939, *The Social Function of Science,* Routledge & Kegan Paul, London.

Black, N., 1997, 'A national Strategy for Research and Development: Lessons from England', *Annu Rev Public Health,* vol 18, pp. 485-505.

Black, D., Pole, R., 1975, 'Priorities in biomedical research', *British Journal of Preventive Social Medicine,* vol. 29, p. 222.

Blau, P., 1964, *Exchange and Power in Social Life,* John Wiley & Sons, Chichester.

Blume, S.S., 1974, *Towards a Political Sociology of Science,* The Free Press, New York.

Blume, S.S., (ed) 1977, *Perspectives in the Sociology of Science,* John Wiley & Sons, Chichester.

Blume, S.S., 1982, *The Commissioning of Social Research by Central Government,* Social Science Research Council, London.

Booth, T.A., (ed) 1979, *Planning for Welfare, Social Policy and the Expenditure Process,* Basil Blackwell and Martin Robertson, Oxford.

Bourdieu, P., 1975, 'The specificity of the scientific field and the social conditions of progress', *Social Sciences Information* vol. 4, no. 6, pp. 19-47.

Braybrooke, D., Lindblom, C.E., 1963, *A Strategy of Decision,* The Free Press, New York.

Brennan, G., Pettit, P., 2004, *The Economy of Esteem*, Oxford University Press, Oxford.

Bridges, E., 1950, *Portrait of a Profession,* Cambridge University Press, Cambridge.

Brown, G.W., Harris, T., 1978, *Social Origins of Depression,* Tavistock Publications, London.

Brown, M., Madge, N., 1982*, Despite the Welfare State,* SSRC-DHSS Studies in Deprivation and Disadvantage, Heinemann, London.

Brown, R.G.S., 1970, *The Administrative Process in Britain,* Methuen, London.

Bulmer, M., (ed) 1978, *Social Policy Research,* Macmillan, London.

Bulmer, M., 1982, *Using Social Science Research in Policy Making. Why are the Obstacles so Formidable?,* Public Administration conference, September, unpublished.

Burns, T., Stalker, C.M., 1966, *The Management of Innovation,* Tavistock Publications, London (2nd ed).

Buxton, M., 1981*, Health Services Research and Health Economics,* memorandum to the Chief Scientist's advisory group of the DHSS from the Health Economics Study Group, unpublished.

Buxton, M.J., Hanney, S., 1996, 'How can payback from health services research be assessed?', *Journal of Health Services Research & Policy*, vol. 1, no. 1, pp. 35-43.

Buxton, M., Hanney, S., Packwood, T., Roberts, S., Youll, P., 2000, 'Assessing Benefits from Department of Health and National Health Service Research & Development', *Public Money and Management*, vol. 20, no. 4, pp. 29-34.

Campbell, D.T., 1979, 'A tribal model of the social system vehicle carrying scientific knowledge', *Knowledge,* December, vol. 1, no. 2.

Campbell, D.T., in Cooke, T., Reichardt, C.S., 1979, *Qualitative and Quantitative Methods in Evaluation Research,* Sage Publications, London/Beverley Hills.

Caplan, N., Morrison, A., Stanbaugh, R.J., 1975, *The Use of Social Science Knowledge. Policy Decisions at the National Level,* Institute of Social Research, University of Michigan, Michigan.

Capra, F., 1975, *The Tao of Physics,* Wildwood House, London.

Central Council for Education and Training in Social Work and Personal Social Services Council, 1980, *Research and practice*, Report of a Working Party on a research strategy for the personal social services.

Cherns, A.A., Sinclair, R., Jenkins, W.I., (eds) 1972, *Social Science and Government: Policies and Problems,* Tavistock Publications, London.

Clark, Burton R., 1983, *The Higher Education System: Academic Organisation in Cross National Perspectives,* University of California Press, Berkeley, CA.

Coombs, R., Richards, A., Saviotti, P., Walsh, V., (eds) 1996, *Technological Collaboration: the Dynamics of Co-operation in Industrial Innovation*, Edward Elgar, Cheltenham.

Crewe, I., (ed) 1974, *British Political Sociology Yearbook,* vol. I, *Elites in Western Democracy,* Introduction, Croom Helm, London.

Croxson, B., Hanney, S., Buxton, M., 2001, 'Routine monitoring of performance: what makes health research and development different?' *Journal of Health Services Research & Policy*, vol. 6, no. 4, pp. 226-32.

Dahl, R., 1971, *Polyarchy, Participation and Opposition,* Yale University Press, New Haven.

Dahlstrom, E., 1979, *Interactions Between Practitioners and Social Scientists in Research and Development into Higher Education,* National Board of Universities and Colleges, Stockholm.

Dash, P., 2003, *Increasing the Impact of Health Services Research on Health Services Improvement,* The Health Foundation and the Nuffield Trust, London.

Davis, P., Howden-Chapman, P., 1996, 'Translating research findings into health policy', *Soc Sci Med* vol. 43, pp. 865-72.

Davies, H.T.O., Nutley, S.M., 2000, Healthcare: evidence to the fore. In: *What Works? – Evidence-based Policy and Practice in Public* Services (Edited by Davies, H.T.O., Nutley, S.M., Smith, P.C.) The Policy Press, Bristol, pp. 43-67.

Davies, H.T.O., Nutley, S.M., Smith, P.C., 2000, *What Works? – Evidence-based Policy and Practice in Public Services*, The Policy Press, Bristol.

Denis, J., Lomas, J., 2003, 'Convergent evolution: the academic and policy roots of collaborative research', *Journal of Health Services Research & Policy,* vol. 8 Suppl. 2, pp.1-6.

DHSS Planning Committee Paper PC (73)/17.29r.

Donnison, D., 1972, 'Research for policy', *Minerva*, September, vol. 10, no. 4, pp. 519- 36.

Downs, A.,1969, *Inside Bureaucracy,* Little, Brown, Boston, MA.

Dunsire, A., 1981, 'Central control over local authorities: a cybernetic approach', *Public Administration,* Summer, vol. 59, pp. 173-88.

Easton, D., 1953, *The Political System,* Alfred A. Knopf, New York.

Easton, D., 1965, A *Framework for Political Analysis,* Prentice-Hall, Engelwood Cliffs, NJ.

Fay, B., 1975, *Social Theory and Political Practice,* George Allen & Unwin, London.

Etzkowitz, H., Leydesdorff, L., 1995, The triple helix–university–industry–government relations: a laboratory for knowledge-based economic development. *EASST Review,* vol. 14, no. 1, pp. 14–19.

Etzkowitz, H., Leydesdorff, L., 2000, 'The dynamics of innovation: from National Systems' and 'Mode 2 to a Triple Helix of university-industry-government relations', *Research Policy*, vol. 29, pp. 109-23.

Ferlie, E., Wood, M., 2003, 'Novel mode of knowledge production? Producers and consumers in health services research', *Journal of Health Services Research & Policy*, vol. 8, Suppl. 2, October, pp.51-7.

Frederiksen, L.F., Hansson, F., Wenneberg, S.B., 2003, 'The *Agora* and the role of research evaluation', *Evaluation*, vol. 9, pp. 149-172.

Frenk, J., 1992, 'Balancing relevance and excellence: organizational responses to link research with decision-making', *Soc Sci Med,* vol. 35, pp.1397-1404.

Fulop, N., Allen, P., Clarke, A., Black, N., 2003, 'From health technology assessment to research on the organisation and delivery of health services: addressing the balance', *Health Policy,* vol. 63, pp. 155-65.

Garrett, J., 1972, *The Management of Government,* Pelican Books, Harmondsworth.

Gibbons, M., Johnston, R., 1974, 'The roles of science in technological innovation' *Research Policy,* vol. 3, pp. 220-42.

Gibbons, M., Gummett, P., 1976, 'Recent changes in the administration of government research and development in Britain', *Public Administration,* Autumn, vol. 54, pp. 247-66.

Gibbons, M., Georghiou, L., 1987, *Evaluation of Research: A Selection of Current Practices*, Paris, OECD.

Gibbons, M., Limoges, C., Nowotny, H., Schwartzman, S., Scott, P., Trow, M., 1994, *The New Production of Knowledge: the dynamics of science and research in contemporary societies*, Sage, London.

Giddens, A., 1974, *Positivism and Sociology,* Heinemann, London.

Glaser, B., Strauss, A., 1967, *The Discovery of Grounded Theory,* Aldine Publishing Co., New York.

Godin, B., (1998) 'Writing Performative History: The New *New Atlantis*?' *Social Studies of Science, vol.* 28, no. 3, pp. 465-483.

Goldberg, E.M., Warburton, R., 1979, *Ends and Means in Social Work,* George Allen and Unwin, London.

Gornitzka, A., (2003) *Science, Clients and the State: A study of scientific knowledge production and use*, CHEPS/ University of Twente, Enschede.

Gordon, M.D., Meadows, A.J., 1981, *The Dissemination of Findings of DHSS Funded Research,* Primary Communications Research Centre, University of Leicester, Leicester.

Graycar, A., 1979, 'Political issues in research and evaluation', *Evaluation Quarterly,* August, vol. 3, no. 3.

Gummett, P., 1980, *Scientists in Whitehall,* Manchester University Press, Manchester.

Habermas, J., *1971, Towards a Rational Society,* Heinemann, London.

Hagstrom, W.O., 1965, *The Scientific Community,* Basic Books, New York.

Haines, A., Kuruvilla, S., Borchert, M., 2004, 'Bridging the implementation gap between knowledge and action for health', *Bulletin of the World Health Organization,* vol. 82, no. 10, pp. 724-32.

Hall, P., Land, H., Parker, RA., Webb, A., *1975, Change, Choice and Conflict in Social Policy,* Heinemann, London.

Hall, P.A., Taylor, R.C.R., 1996, 'Political Science and the Three New Institutionalisms'. *Political Studies*, vol. XLIV, pp. 936-957.

Ham, C., 1981, *Policy Making in the National Health Service,* Macmillan, London.

Hammond, K.R., Mumpower, J., 1979, 'Formation of social policy: risks and safeguards', *Knowledge,* December, vol. 2, no. 2.

Hanney, S., Packwood, T., Buxton, M., 2000, Evaluating the benefits from health R&D centres: a categorisation, a model and examples of application. *Evaluation: the International Journal of Theory, Research and Practice,* vol. 6, no.2, pp. 137-60.

Hanney, S.R., Gonzalez-Block, M.A., Buxton, M.J., Kogan, M., 2003, 'The utilisation of health research in policy-making: concepts, examples and methods of assessment', *Health Research Policy and Systems*, vol 1, no. 2. [Available at: http://www.health-policy-systems.com/content/pdf/1478-4505-1-2.pdf]

Hanney, S., Soper, B., Buxton, M., 2003a, *Evaluation of the NHS R&D Implementation Methods Programme. HERG Research Report No 29,* Health Economics Research Group, Brunel University, Uxbridge. [Available at: http://www.brunel.ac.uk/depts/herg/pubs/internal.html]

Harris, M., 1982, *Report on a Postal Survey of Salaried Researchers.*

Harrison, A., New, B., 2002, *Public Interest, Private Decisions: Health-related research in the UK,* King's Fund, London.

Health Foundation, 2003, *Increasing the Impact of Health Services Research on Service Delivery: A Consultation Document,* The Health Foundation, London.

Heady, J.A., [n.d.], *To Investigate and to Publish,* Joseph Rowntree Memorial Trust, York.

Henkel, M., 1994, 'PSSRU Research on Care Management: Kent Community Care Project' in Buxton, M., Elliott, R., Hanney, S., Henkel, M., Keen, J., Sculpher, M. and Youll, P., 1994, *Assessing Payback from Department of Health Research and Development: Preliminary Report*, Vol. 2: *Eight Case Studies*, HERG Research Report no. 19, Health Economics Research Group, Brunel University, Uxbridge.

Henkel, M., 2000, *Academic Identities and Policy Change in Higher Education*, Jessica Kingsley Publishers, London.

Henkel, M., 2005, 'Academic Identity and Autonomy in a changing policy environment', *Higher Education*, Vol. 44, No.1, pp. 155-77.

Henkel, M., Kogan, M., 1981, *The DHSS Funded Research Units: The Process of Review*, Department of Government, Brunel University, Uxbridge.

Henkel, M., Hanney, S., Kogan, M., Vaux, J., von Walden Laing, D., 2000, *Academic Responses to the UK Foresight Programme,* Centre for the Evaluation of Public Policy and Practice, Brunel University, Uxbridge.

Hill, M., 1979, 'Implementation of the central-local relationship', in *Central Local Government. Relationships. A Panel Report to the Research Initiatives Board,* Social Science Research Council, London.

Husén, T., Kogan, M., 1984, *Researchers and Policy Makers in Education. How do they relate?,* Pergamon Press, Oxford.

Innvær, S., Vist G., Trommald M., Oxman, A., 2002, 'Health policy-makers' perceptions of their use of evidence: a systematic review'. *Journal of Health Services Research & Policy* 2002, vol. 7, pp. 239-44.

INVOLVE, 2004, *About INVOLVE,* [http://www.invo.org.uk/About_Us.asp Accessed 24 Feb 2005].

Irvine, J., Martin, B., 1984, *Foresight in Science: Picking the Winners,* Pinter Publishers, London.

Irwin, A., 2001, 'Constructing the scientific citizen: science and democracy in the biosciences', 2001, *Public Understanding of Science,* vol. 10, pp. 1-18.

Jones, R.E., 1966, *The Functional Analysis of Politics,* Routledge & Kegan Paul, London.

Keeling, D., 1972, *Management in Government,* George Allen & Unwin, London.

Kingdon, D.R., 1973, *Matrix Organisation:* Managing Information Technology, Tavistock Publications, London.

Knox, E.G., 1980, *Research and Development in Health Services,* European Symposium on Medical Statistics.

Kogan, M., 1969, *The Government of the Social Services,* Charles Russell Memorial Lecture.

Kogan, M., with Boyle E., Crosland, A., 1971, *The Politics of Education,* Penguin, Harmondsworth.

Kogan, M., et al, 1978, Research Report, No.1 Royal Commission on the National Health Service, *The Working of the National Health Service*, HMSO, London.

Kogan, M., Korman, N., 1975, 'DHSS and field authorities', July, and Korman, N., 1976, Relationships between the DHSS and SSDs in setting research objectives', May (unpublished project papers).

Kogan, M., Korman, N., Henkel, M.,1980, *Government's Commissioning of Research. A Case Study,* Brunel University, Uxbridge.

Kogan, M., Hanney, S., 2000 *Reforming Higher Education,* Jessica Kingsley, London.

Kogan, M., Henkel, M., 1983, *Government and Research: The Rothschild Experiment in a Government Department,* Heinemann, London.

Kogan, M., Henkel, M., 2000, 'Future Directions for Higher Education Policy. Getting Inside: Policy Reception of Research', in Schwartz, S and Teichler, U., (ed) *The Institutional Basis of Higher Education Research*, Kluwer, Dordrecht, Holland.

Korman, N., 1976, The By-pass Procedure of the Small Grants Scheme. (Unpublished. BRMP (76) /3).

Korman, N., Kogan, M., 1978, *Planning System in a Region: A Case Study, Royal Commission on the National Health Service, Research Paper No.* 1, HMSO, London.

Krohn, R., 1977, 'Scientific ideology and scientific process: the natural history of a conceptual shift', in Mendelsohn, E., Weingart P., Whitley R., (eds), *The Social Production of Scientific Knowledge, vol.* 1, D. Reidel Publishing Co., Dordrecht, Holland/Boston, USA.

Kuhlmann, S., 2001, 'Future governance of innovation policy in Europe – three scenarios', *Research Policy,* vol. 30, pp. 953-976.

Kuhn, T.S., 1969, 'The history of science', in *International Encyclopaedia of the Social Sciences 14,* pp. 74-83.

Kuhn, T.S., 1962, *The Structure of Scientific Revolutions,* University of Chicago Press, (2nd ed,1970), Chicago, Ill.

Kuhn, T.S., 1974, 'Logic of discovery or psychology of research?' in Lakatos, I., Musgrave, A., (eds), *Criticism and the Growth of Knowledge,* Cambridge University Press, Cambridge.

Lambert, R., 1963, *Sir John Simon 1816- 1904,* McGibbon & Kee, St. Albans, Herts.

Latour, B., 1987, *Science in Action: how to follow scientists and engineers through society*, Harvard University Press, Cambridge MA.

Latour, B., Woolgar, S., 1986, *Laboratory Life: the social construction of scientific facts*, Harvard University Press, Cambridge MA.

Lavis, J.N., Ross, S.E., Hurley, J.E., Hohenadel, J.M., Stoddart, G.L., Woodward, C.A., Abelson, J., 2002, 'Examining the role of health services research in public policymaking'. *Milbank Q*, vol. 80, pp.125-54.

Lavis, J., Posado, F., Haines, A., Osei, E., 2004, 'Use of research to inform public policymaking', *The Lancet,* vol 364, 1615-21.

Lessnoff, M., 1973, *The Structure of Social Science,* George Allen & Unwin, London.

Lomas, J., 1997, *Improving Research Dissemination and Uptake in the Health Sector: Beyond the Sound of One Hand Clapping*, Centre for Health Economics and Policy Analysis, McMaster University, Hamilton.

Lomas, J., 2000, 'Using 'linkage and exchange' to move research into policy at a Canadian Foundation'. *Health Aff*, vol. 19, 236-40.

Lynch, M., 1993, *Scientific Practice and Ordinary Action: Ethnomethodology and Social Studies of Science*, Cambridge University Press, Cambridge.

Lindblom, C.E., Cohen, D.K., 1979, *Usable Knowledge,* Yale University Press, New Haven.

Linder, S.H., 1980, *Perceptions of the Policy Making Environment* (unpublished).

Macintyre, S., Chalmers, I., Horton, R., Smith R., 2001, 'Using evidence to inform health policy: case study', *BMJ,* vol. 322, pp. 222-5.

McLachan, G., (ed), 1978, *Five Years After: A review of Health Care Research Management After Rothschild,* Nuffield Provincial Hospitals Trust, London.

McLachan, G., 1985, *A Fresh Look at Policies for Health Services Research and its Relevance to Management,* Nuffield Provincial Hospitals Trust, London.

McLeod, R., 1977, 'Changing perspectives in the social history of science', in Spiegel- Rösing, 1. and de Solla Price, D., *Science, Technology and Society,* Sage Publications, London/Beverley Hills.

McNally, N., Kerrison, S., Pollock, A., 2003, 'Reforming clinical research and development in England', *BMJ,* vol. 327, pp. 550-3.

Magee, B., 1973, *Popper,* Fontana Modern Masters.

Marin, B., 1981, 'What is "half-knowledge" sufficient for - and when? Theoretical comment on policymakers' uses of social science', *Knowledge,* vol. 3, no. 1, pp. 43-60.

Marin, B., Mayntz, R., 1991, *Policy Networks. Empirical Evidence and Theoretical Considerations,* Campus, Frankfurt, Westview Press, Boulder.

Marris, P., Rein, M., 1974, *Dilemmas of Social Reform,* Penguin, Harmondsworth, (2nd ed).

Martin, B., 1996, 'Technology Foresight: Capturing the Benefits from Science Related Technologies' *Research Evaluation,* vol 6, no. 2, pp. 158-68.

Martin, M., Nightingale, P., 2000, *The Political Economy of Science, Technology and Innovation*, Edward Elgar, Cheltenham.

Maynard, A., Bloor, A., Freemantle, N., 2004, 'Challenges for the National Institute for Clinical Excellence', *BMJ,* vol. 329, pp. 227-9.

Mays, N., Roberts, E., Popay, J., 2001, 'Synthesising research evidence', in Fulop, N., Allen, P., Clarke, A., Black, N., (eds) *Studying the Organisation and Delivery of Health Services: Research Methods,* Routledge, London.

Mendelsohn, E., 1973, 'A human reconstruction of science', *Boston University Journal,* Spring.

Mendelsohn, E., Weingart, P., Whitley, R., (eds), 1977, *The Social Production of Scientific Knowledge, vol.* 1, D. Reidel Publishing Co., Dordrecht, Holland/Boston, USA.

Merton, R.K., 1973, *The Sociology of Science,* Chicago University Press, Chicago, Ill.

Milewa, T., Buxton, M., Hanney, S., 2005, 'Lay involvement in the public funding of medical research: expertise and counter-expertise in empirical and analytical perspective', *Critical Public Health.*

Minerva 1972, 'The choice and formulation of research problems: four comments on the Rothschild Report', vol. 10, no. 2, pp. 191- 208.

Mitroff, I., 1974, 'Norms and Counternorms in a Select Group of the Apollo Moon Scientists: a case study of the ambivalence of scientists', *American Sociological Review,* vol. 39, pp. 570-95.

Molas-Gallart J., Tang, P., Morrow, S., 2000, 'Assessing the non-academic impact of grant-funded socio-economic research: results from a pilot study', *Research Evaluation,* vol 9, no.3, pp. 171-82.

Moss, L., 1977, *Some Attitudes Towards Research,* (unpublished).

Mulkay, M.J., 1972, *The Social Process of Innovation: A Study in the Sociology of Science,* Macmillan, London.

Mulkay, M.J., 1977, 'Sociology of the scientific research community', in Spiegel-Rösing, I, de Solla Price, D., *Science, Technology and Society,* Sage Publications, London/ Beverley Hills.

Mulkay, M.J., 1979, *Science and the Sociology of Knowledge,* George Allen & Unwin, London.

Mussachia, M.M., 1979, *Scientific Development: Theoretical issues and Perspectives,* University of Gothenburg, Gothenburg, Sweden.

Nagel, E., 1961, *The Structure of Science,* Routledge & Kegan Paul, London.

Nisbet, J., Broadfoot, P., *1980, The Impact of Research on Policy and Practice in Education,* Aberdeen University Press, Aberdeen.

Nodder, T., 1979, talk given to graduate weekend meeting, Brunel University, unpublished.

Nowotny, H., Scott, P., Gibbons, M., 2001, *Re-thinking Science: Knowledge and the Public in an Age of Uncertainty,* Polity, Cambridge.

O'Donnell, M., Entwistle V., 2004, 'Consumer involvement in decisions about what health-related research is funded', *Health Policy,* vol. 70, pp. 281-90.

OECD, 1971, *Science, Growth and Society: a New Perspective.* Report of the Secretary-General's Ad Hoc Group on New Concepts of Science Policy, chaired by Harvey Brooks, Paris: OECD

OECD, 1975, *Educational Development Strategy in England and Wales,* Paris.

Oliver, S., Clarke-Jones, L., Rees, R., Milne, R., Buchanan, P., Gabbay, J., Oakley, A., Stein, K., 2004, 'Involving consumers in research and development agenda setting for the NHS: developing an evidence-based approach', *Health Technol Assess,* vol. 8, no. 15.

Pang, T., Sadana, R., Hanney S., Bhutta Z.A., Hyder, A.A., Simon, J., 2003, 'Knowledge for better health – a conceptual framework and foundation for health research systems', *Bulletin of the World Health Organization*, vol. 81, no. 11, pp. 815-20.

Pantin, C., 1968, *The Relations between the Sciences*, Cambridge University Press, Cambridge.

Pavitt, K., 1991, 'What makes basic research economically useful?', *Research Policy, vol.* 20, pp.109-19.

Peckham, M., 1991, 'Research and development for the National Health Service', *The Lancet,* vol 338, pp. 367-71.

Peckham, M., 1999, 'Developing the National Health Service: a model for public services', *The Lancet, vol.* 352, pp.1539-45.

Perry, N., 1975, *The Organisation of Social Science Research in the UK,* Social Science Research Council Survey Unit, London.

Pile, Sir William, 1976, evidence to Fookes Committee.

Platt, J., 1976, *Realities of Social Research,* Sussex University Press, Brighton.

Platt, J., 1987, 'Research dissemination: a case study', *The Quarterly Journal of Social Affairs,* vol. 3, pp. 181-98.

Pleasant, A., Kuruvilla, S., Zaracadoolas, C., Shanahan, J., Lewenstein, B., 2003, *A Framework for Assessing Public Engagement with Health Research*, Available at: www.cook.rutgers.edu/~CILS/Pleasant_etal_WHO.pdf

Polanyi, M., 1958, *Personal Knowledge,* Chicago University Press, Chicago, Ill.

Polanyi, M., 1962, 'The republic of science: its political and economic theory', *Minerva,* Autumn, vol. 1, no. 1. pp. 54-73.

Pollitt, C., 1993, 1995, *Managerialism and the Public Services. The Anglo-American Experience,* (1st 2nd edns,) Blackwell, Oxford.

Pollitt, C., Hanney, S., Packwood, T., Rothwell, S. and Roberts, S., 1997, *Trajectories and Options: An International Perspective on the Implementation of Finnish Public Management Reforms,* Ministry of Finance, Helsinki.

Popper, K.R., 1966, *The Open Society and its Enemies,* Routledge & Kegan Paul, London, (5th ed).

Popper, K.R., 1972, *The Logic of Scientific Discovery,* Hutchinson, London, (rev. ed).

Premfors, R., 1979, *Social Research and Public Policy Making: An Overview,* Yale Higher Education Research Group Working Paper.

Premfors, R., 1981, Review article, Charles Lindblom and Aaron Wildavsky, *British Journal of Political Science,* April, vol. 11, part 2.

Price, D.K., 1965, *The Scientific Estate,* Harvard University Press, Cambridge, MA.

Ranson, S., 1980, 'Changing relations between centre and locality in education', *Local Government Studies,* vol. 6, no. 6.

Rein, M., 1976, *Social Science and Public Policy,* Penguin, Harmondsworth.

Rein, M., White, S.H., 1977, 'Policy Research: Belief and Doubt', *Policy Analysis,* vol. 3.

Rein, M., White, S.H., 1980, *Knowledge for Practice: The Study of Knowledge in Context for the Practice of Social Work,* MIT.

Rhodes, R.A.W., 1979, *Research into Central-Local Relationships in Britain: A Framework for Analysis,* Social Science Research Council. Report on central-local government relationships, Appendix 1, London.

Richardson, A., Jackson, C., Sykes, W., 1990, *Taking Research Seriously,* Department of Health, London.

Richardson, J.J., Jordan, A.G., 1979, *Governing Under Pressure, The Policy Process in a Post-Parliamentary Democracy,* Martin Robertson, Oxford.

Rip, A., 2000, 'Fashions, lock-ins and the heterogeneity of knowledge production' in M. Jacob and T. Hellström (eds) *The Future of Knowledge Production in the Academy*, Open University Press, Buckingham.

Rivlin, A.M., 1971, *Systematic Thinking for Social Action,* The Brookings Institution, Washington DC.

Rose, H., Rose S., 1976, *The Political Economy of Science,* Macmillan, London.

Rose, R., 1977, 'Disciplined research and undisciplined problems' in Weiss, C *Using Social Research in Public Policy Making,* D.C. Heath, Lexington, MA.

Ruin, O., in T. Husén., M. Kogan, 1984, *Researchers and Policy Makers in Education. How Do They* Relate?' Pergamon Press, Oxford.

Salomon, J.J., 1977, 'Science policy studies in the development of science policy', in Spiegel-Rösing, I., de Solla Price, D., *Science, Technology and Society,* Sage Publications, London/Beverley Hills.

Salter, B., 1982, *Contract Research: Universities and the Knowledge Market,* University of Surrey, Guildford (unpublished).

Schon, D., 1971, *Beyond the Stable State,* Random House, New York.

Schonfield, A., 1972, 'The social sciences in the great debate on science policy', *Minerva,* July, vol. 10, no. 3, pp. 426-38.

Schutz, A., 1972, *The Phenomenology of the Social World,* Heinemann, London.

Self, P., 1972, *Administrative Theories and Politics,* George Allen & Unwin, London.

Shepherd, M., 1982, 'Psychiatric research in medical perspective', *British Medical Journal,* March, pp. 961-3.

Shils, E., 1982, 'Knowledge and the sociology of knowledge', *Knowledge,* September, vol. 4, no. 1.

Simon, H., 1945, *Administrative Behaviour*, Macmillan, London.

Smith, C., 1982, *The Changing Social Organisation of the Social Sciences in Britain and its Implication for Intellectual Orthodoxies,* (unpublished).
Smith, R., 1988, 'A national health research policy', *BMJ,* vol. 297, pp. 805-806.

Social Research Association, 1980, *Terms and Conditions of Social Research Funding in Britain.*

Social Research Association, 1981, *Social Research and Public Policy. Three Perspectives.*

Social Science Research Council, 1977, *Health and Health Policy,* London.

Spiegel- Rösing, I., de Solla Price, D., 1977, *Science, Technology and Society,* Sage Publications, London/Beverley Hills.

Stacey, M., 1977, 'Concepts of health and illness: a working paper on the concepts and their relevance for research.' Appendix 3 , *Health and Health Policy,* SSRC, London.

Subramaniam, V., 1963, 'Specialists in British and Australian government services, a study in contrasts', *Public Administration,* pp. 357-73.

Task Force on Health Systems Research. 2004, 'Informed choices for attaining the Millenium Development Goals: towards an international cooperative agenda for health systems research', *The Lancet,* vol. 364, pp. 997-1003.

Taylor, D., 1981, *Health Research in England. A Topic for Debate,* Office of Health Economics, London.

The NHS Confederation, 2005, *The Health Services Research Network,* [http://www.nhsconfed.org/influencing/health_services_research_network.asp Accessed 21 Jan 2005].

Thomas, P., 1982, 'Social research and government policy, Heyworth, Rothschild and after', *Futures,* February.

Trist, E., 1972, 'Types of output mix of research organisations and their complementarity', in Cherns A.B. *et al., Social Science and Government: Policies and Problems,* Tavistock Publications, London.

Trostle, J., Bronfman, M., Langer, A., 1999, How do researchers influence decision-makers? Case studies of Mexican policies. *Health Policy Plan*, vol. 14, pp. 103-14.

Trow, M., 1980, 'Researchers, policy analysts and policy intellectuals , in Husén, T., Kogan, M., 1984, *Researchers and Policy Makers in Education. How do they relate?,* Pergamon Press, Oxford.

UK Clinical Research Collaboration, 2004, The UKCRC Partners, UKCRC, London [http://ukcrc.org/UKCRC_Partners.htm Accessed 24 Feb 2005].

University of Aberdeen, Department of Community Medicine, 1980, *Economists in Multidisciplinary Teams: Some Unresolved Problems in the Conduct of Health Services Research,* Aberdeen.

Van den Daele, W., Krohn, W., Weingart, P., 1977, 'The political direction of scientific development', in Mendelsohn, E., Weingart, P., Whitley, R., (eds)*, The Social Production of Scientific Knowledge, vol.* 1, D. Reidel Publishing Co., Dordrecht, Holland/Boston, MA.

Van der Meulen, B., Rip, A., 2000, 'Evaluation of societal quality of public sector research in the Netherlands', *Research Evaluation,* vol. 9, no. 1, pp. 11-25.

Walt, G., 1994, *Health Policy: An Introduction to Process and Power,* Zed Books, London.

Walter, I., Davies, H., Nutley, S., 2003, 'Increasing research impact through partnerships: evidence from outside health care', *Journal of Health Services Research & Policy,* vol. 8, Suppl. 2, October, pp. 58-61.

Watson, J.D., 1968, *The Double Helix,* Weidenfeld & Nicolson, London.

Weale, A., 2001, 'Science advice, democratic responsiveness and public policy', *Science and Public Policy,* vol. 28, no. 6, pp. 413-21.

Weingart, P., 1974, 'On a sociological theory of scientific change', in Whitley, R.D., (ed), *Social Processes of Scientific Development,* Routledge & Kegan Paul, London.

Weingart, P., 1977, 'Science policy and the development of science', in Blume, S., (ed), *Perspectives in the Sociology of Science,* J. Wiley & Son, Chichester.

Weingart, P., 1997, 'From Finalisation to Mode 2: old wine in new bottles?', *Social Science Information,* vol. 36, no. 4, pp. 591-613.

Weiss, C., Bucuvalas, M.J., 1977, 'The challenge of social research to decision making', in Weiss, C., *Using Social Research in Public Policy Making,* D.C. Heath, Lexington, MA.

Weiss, C.H., 1977, *Using Social Research in Public Policy Making,* D.C. Heath, Lexington, MA.

Weiss, C.H., 1980, *Social Science Research and Decision Making,* Columbia University Press, Newark.

Whitley, R.D., 1977, 'Changes in the social and intellectual organisation of the sciences- professionalisation and the arithmetic ideal', in Mendelsohn, E., Weingart, P., and Whitley, R., (eds), *The Social Production of Scientific Knowledge, vol.* 1, D. Reidel Publishing Co., Dordrecht, Holland/Boston, MA.

Wildavsky, A., 1964, *The Politics of the Budgetary Process,* Little, Brown, Boston, MA.

Wildavsky, A., 1966, 'The political economy of efficiency: cost-benefit analysis, systems analysis and program budgetting', *Public Administration Review,* vol. XXVI, pp. 292-310.

Wildavsky, A., 1979, *Speaking Truth to Power,* Little, Brown, Boston, MA.

Williams, A., 1997, *Being Reasonable about the Economics of Health: Selected Essays by Alan Williams.* Compiled and edited by Culyer, A.J. and Maynard, A., Edward Elgar, Cheltenham.

Williams, A., 1981, 'Economics and health services research', Appendix 1, in Taylor, D., *Health Research in England: A Topic for Debate,* Institute of Health Economics.

Williams, W., 1971, *Social Policy Research and Analysis, The Experience of Federal Social Agencies,* Elsevier, Holland.

Winch, P., 1958, *The idea of a Social Science and its Relations to Philosophy,* Routledge & Kegan Paul, London.

Wittrock, B., Lindstrom, S., Zetterbey, K., 1981, *Implementation Beyond Hierarchy. Swedish Energy Research Policy,* University of Stockholm, Group for Advanced Study of Higher Education and Research Policy, Sweden.

Yin, R.K., Gwaltney, M.K., 1981, 'R & D management practices.- knowledge utilization as a networking process', *Knowledge,* June, vol. 2, no. 4, pp. 555-80.

Ziman, J.M., 1968, *Public Knowledge,* Cambridge University Press, Cambridge.

Ziman, J.M., 1981, 'Conceptions of science', conference on the political realisation of social science knowledge and research, Vienna.

Index

accountability, 14, 41, 45, 56, 62, 118, 152-154, 170, 200
  research management, 17, 55, 165 *bis*
  research unit, 98, 124, 126, 131
  scientific, 20, 103, 131, 135, 169, 181
action research, 34, 100, 111, 124, 139, 145
  and service developments, 100, 101
addictions (RLG), 95, 97, 98, 102, 103
administrators, 17, 19, 56-59, 178, 179
  generalist, 44, 45, 183, 184
adolescents, disturbed (RLC), 103, 104
adversarial mode, 138, 148-151
Advisory Board for Research Councils, 87
Agricultural Research Council, 15
Alliance for Health Policy and Systems Research (AHPSR), 3
applied research, 14, 16, 28, 31, 32, 36, 86, 106, 119, 125
  fundamental and, 14, 16, 28, 31, 32, 36, 106, 119, 125
  steerage of, *see* steerage
authority
  of government, 151, 152
  of science, 24, 35, 108, 131, 160, 161, 206
  -power relationship, 1, 8, 150

behaviour studies, 34, 121
Better Services for the Mentally Handicapped (1971), 219
Better Services for the Mentally Ill (1975), 219
biomedical research, 63, 77, 80, 83-86, 88-90, 120, 206
  collaboration attempt, 80
  MRC role and policy, 63, 77, 79, 80
  PMR role, 73, 78-80, 171
boundary issues, 34, 94
  liaison officer role, 130
  multi-modality and, 5, 198, 202, 203
  RLG solution, 94
brokerage roles, 20, 42, 103, 187, 209, 211
  Chief Scientist, 3, 12, 20, 89, 177, 182-184
  liaison officers, 20, 186, 187
bypass mechanism (SGC), 109

Canadian Health Services Research Foundation, 3
case register (RLG resource), 100, 172
central government, *see* government
Central Policy Review Staff, 44
Central Research and Development Committee (CRDC), 194
Centre for Reviews and Dissemination, 195
Centre for Studies in Social Policy, 102
Chief Scientist, 10, 15, 17-21, 44, 45, 51-53, 61-63, 68, 72-74, 77, 79, 82, 87, 89, 90, 93, 97, 104, 109, 115, 116, 118, 129-133, 135-137, 141-145, 148-151, 153, 155, 157-162, 170, 177, 180-184, 186, 192, 209, 213, 214, 216
  brokerage by, 3, 12, 20, 89, 177, 182-184, 209
  Committees, 53, 61, 62, 182
  roles, 17, 20, 68, 103, 112, 153
Chief Scientist's visits, 104, 129-132, 135-137, 141-145, 148-151, 153-155, 157, 161, 162, 214
  alternative models, 151
  arrangements for, 132
  customer review, 143, 161 *passim*
  procedures, 132, 143
  project presentation, 129

report stage, 142, 154
scientific merit, 131, 137, 144, 145, 157, 161
Chief Scientist's Organisation, 76, 77, 87, 109, 115
Chief Scientist's Research Committee (CSRC), 17-21, 48, 52, 53, 61-73, 75, 76, 80, 84, 93, 95, 101-103, 105, 112, 168, 169, 171, 180, 182, 185, 196, 205, 213
disbandment, 75
evaluative research, 72, 101, 112
intermediate levels and, 73
macro-scientific policy, 53, 61, 71, 72
membership, 61, 64, 65
policy priorities, 62, 73
RLG guidance, 63, 65, 174
role, 20
children, 55, 57, 93, 96, 103, 104, 123, 147, 195
RLG projects, 93-98, 103, 107 *passim*
Civil Service, 44, 57, 59
client groups, 8, 9, 11, 13, 43, 45, 55, 57, 58, 63, 74, 75, 93, 95-97, 100, 103, 127, 155, 173, 174, 178, 203, 207
impact of research, 173
McKinsey Report, 57, 74
Service Development Group, 57, 58, 74, 178
clinical trials, 80
Cochrane Collaboration, 195, 206
cognitive structures, 120, 191 *passim*
of science, 26, 191
coherence concept, 118
collaboration model, 209
collaborative approach, 3, 197, 210
Collaborative Training Programme (CTP) 3
communism, 24
communication, knowledge (control), 26
community care, 55, 74, 88, 96, 103
community health councils, 40, 55
confidentiality concept, 70, 71, 118
configurational sciences, 26 *bis*
control concept, 23, 97, 107, 118
Controller Research and Development (DHSS), 21
convergence and divergence in government and science, 8-10, 202
cost-benefit analysis (CBA), 43, 44, 68, 69
Council on Health Research for Development (COHRED), 3
curriculum building project, 139
customer, 3, 5, 9, 11, 13-17, 19, 28, 57-59, 69, 73, 75-77, 82-83, 86, 89, 93, 96-98, 104, 110, 117, 119, 124, 131, 132, 153-162, 165-170, 173, 174, 177-182, 185, 187-189, 191-201, 203, 204, 209-211, 213, 215
primary, and secondary or proxy roles, 175, 177, 178, 193, 194, 203
review (units), 5, 129, 143, 152, 153, 156-158, 165
roles, 5, 68, 153, 154, 177-180, 192, 197, 198, 203, 205, 210
customer-contractor principle
applied research, 11, 14, 15
exchange theory, 11, 204
health department role, 90
in Medical Research Council, 76, 90
RLGs' role, 19, 180
systematised collaboration, 19
Dainton Report (1971), 14, 24
decision-making, 11, 19, 34, 40, 45, 46, 50, 55, 58, 85, 126, 154, 155, 160, 188
Department of Education and Science (DES), 77, 89
Department of Employment, 103
Department of the Environment, 103
Department of Health (DH) 3, 89, 196, 198, 220, 221, 224, 228, 233

Department of Health and Social Security (DHSS), 1, 4, 5, 7, 9-13, 16-21, 23, 30, 34, 36, 39, 41-45, 47, 48, 50, 52-59, 61, 63, 664, 67, 70, 73-80, 82, 85, 87, 92, 95, 98, 102-106, 108, 109, 111, 115-121, 124, 126, 127, 129-132, 134, 135, 137, 140-142, 145, 147, 153, 155, 158, 161, 163, 166-170, 172, 173, 175, 177, 178, 180-193, 196-198, 205-207, 209, 213-217
customer roles, 153, 177, 178, 180, 192, 197, 198
Economic Adviser's Office (EAO), 94
exchange relationships, 52, 206
multi-modality, 5
participation/interaction in, 39-41
planning capacity, 42
policy system (HPSS), 124
programme budgeting, 43
research impact, 175
science theories and, 23-29 *passim*
social care seminar, 67, 102
Department of Health and Social Security commissioning system, 17, 50, 53, 59, 61, 111, 185, 213, 214
monitoring, 169, 170
policy system (HPSS), 124
research needs, 103, 177 *passim*
structure, 17
value (analysis), 202
see also Chief Scientist entries; Health Services Research Board; Personal Social Services Research Group; research council system; Research Liaison Groups, research management system; research units; Small Grants Committee
Department of R&D, 194, 198
deprivation, transmitted, 87
Director of R&D, 194, 198
disabled, 96, 123, 147
discipline(s)
-based research, 31, 122
development phases, 29, 30
see also multidisciplinary research
diseases (research priorities), 81, 82, 84, 197
dissemination, 34, 66, 99, 100, 102, 105, 106, 108, 117, 125, 131, 138, 140, 146, 170-172, 175, 177, 192, 192, 195, 200, 203, 207, 215
domain-based research, 31, 35, 122, 147, 151
Down's syndrome babies (project), 31

Economic Adviser's Office, 94
Economic and Social Research Council (ESRC), 207
elderly (projects), 93, 96 *passim*
elites, 9, 11, 26, 27, 47
loosely coupled, 12, 47, 51
employment issues, 103, 140
enlightenment model, 35, 91
environment, policy-makers and, 50
epidemiology, 30, 90, 100, 120-122, 125, 126, 141, 146, 147
ethnic minority groups, 103
evaluative modes/structures, 151
adversarial, 148, 151
alternatives, 151
scientific merit and, 148
evaluative research, 33, 72, 100, 101, 112
CSRC policy, 72, 101
RLG policy, 72
evidence-based policy(ies), 2, 49, 225
exchange relationships, 7, 11, 52, 91, 106, 107, 191, 206
at RLG level, 106
imbalance in, 206
externalist concept of science, 27, 36

family care, 96, 121
feedback, 28, 34, 35, 43, 46, 50, 141, 171
field authorities, 19, 55-59, 94, 101, 172, 174, 175, 177, 178, 188, 189, 209
finalisation, 29
forensic psychiatry (RLG), 93, 103, 119
foresight policies, 2
Fulton Report (1968), 44
funds/finding, 10, 15, 19, 52, 66, 67, 69, 71, 73, 75, 77, 78, 80, 84, 87, 89, 93, 101, 115, 127, 167, 170, 192, 196, 198, 204, 206
  allocation of CSRC, 67
  DHSS, *see* research units
  MRC, 10, 19, 20, 63, 76, 82, 182, 192

generalisability, 101, 105, 106, 134, 157
generalisations for policy-making, 191-211
generalists and specialists, 45, 184
government
  applied research (funding), 14-16
  Central Policy Review Staff, 44
  differentiation within, 45
  exchange relationships, 7, 11, 52, 91, 106, 107, 191, 206
  model (idealised), 7, 8
  multi-modality, 5, 198, 202, 203
  neutrality, 47
  planning (central), 20, 41, 67
  policy-making, 15, 46, 52, 64
  primary/secondary customer, 150-51
  rationality, 26-7, 29-31
  role changes, 26-8
  science and (models), 169-73
  steerage, 19-21, 25, 163, 167, 170
grant applications (SGC), 90-94

Haldane Report (1918), 57
Handbook on Research and Development (DHSS, 1979), 116
health departments, 19, 64, 77-80, 82-84, 86, 89, 90, 192
Health and Personal Social Services (HPSS), 17, 19, 20, 52, 61, 64, 74, 75, 97, 105, 115, 120-123, 127, 131, 134, 146, 161, 177, 178, 182
  research strategy, 182
health services, 19, 20, 29, 30, 35, 55, 85, 87, 90, 109, 110, 118, 120, 166
  planning theories, 41, 42
  research, 20, 29, 30, 35, 89, 90, 110, 115, 118-120, 126, 132, 182, 192, 193, 197, 213
  unified, 55
  see also National Health Service
Health Services Research Board (HSRB), 19, 61, 73
Health Services Research Network, 198, 235
Health Technology Assessment (HTA), 198, 199, 206, 210
Hester Adrian Research Centre, 31
higher education, 51, 91, 120, 204
Home Office, 103, 115
homelessness (RLG), 93, 95, 97, 98, 102, 103
House of Lords Select Committee on Science and Technology, 193-196, 200, 220-222
hypothermia (RLG project), 96

illness, 33, 45, 63, 67, 74, 80, 82, 93, 95, 99, 100, 102-105, 107, 121-123, 132, 139, 169
impact (of research), 34, 165-175
individualist interaction, 50
information sources, 100
  see also knowledge
interaction, 1, 3, 14, 26, 35, 39, 40, 50, 82, 90, 93, 104, 107, 165, 188, 208, 210
interactive model (impact), 35, 175
internalist concept, 7, 8, 23, 25, 36
  externalist and, 28 *passim*

intervention, 33, 34, 36, 100, 121, 122, 200, 208
growth of rationality, 40
intuition (government role), 39

judgements, 9, 24, 40, 41, 56, 59, 67, 68, 71, 85, 97, 106, 123, 133-137, 140-150, 154, 156-163, 171, 182, 186, 187, 213
scientific, 24, 71, 76, 85, 144, 158, 161, 185
knowledge, 2, 4, 7-15, 23-29, 31-33, 35, 40, 45, 57, 67, 72, 73, 75, 79, 83, 85, 88, 90, 93, 100, 111, 115, 121-124, 133, 135, 144, 148, 149, 160, 162, 166, 178, 183, 187, 195, 199, 200, 202-204, 206-210
policy analysis and, 42
scientific, 14, 23, 25, 27, 28, 73, 162
systems (interactive models), 7, 51, 148, 160, 162
knowledge society, 2
knowledge transfer, 35, 203

liaison officers (DHSS), 20, 118, 129-133, 142, 143, 153-157, 167, 170, 171, 177, 186-188, 205, 209
linkage and exchange model, 3, 230
local authorities, 34, 106, 188, 214
social services, 55, 93, 100, 132
Local Government Commissioners, 41

McKinsey Report, 57, 74
macro-scientific policy, 12, 53, 61, 71, 72
managerial model, 184
Medical Research Council (MRC), 10, 15, 16, 19-21, 24, 62, 63, 73, 75-91, 109, 116, 117, 120, 125, 133, 182, 192, 193, 196, 198, 200, 204, 206, 207, 215
DHSS and, 10, 16, 77, 89
funds, 10, 19, 20, 63, 76, 82, 182, 192
health departments and, 89
PMR and, 19, 73, 77, 78, 83, 85, 86
Public Accounts Committee (PAC), 24, 84, 116, 125
medical sociology, 120
mental health, 55, 146, 150, 155, 194
ministers, 9, 43-47, 51, 56, 57, 59, 86, 88, 96
mission-oriented research, 34
Mode 1 and Mode 2 knowledge production, 29
molecular biology, 80, 81
monitoring (projects), 169, 170, 188
multi-modality, 5, 198, 202, 203
multi-level, multi-actor approaches to policy making 17, 49, 203
multidisciplinary research, 68, 71, 99, 112, 147
balance in, 61, 73, 119
CSRC policy, 68, 71, 72
discipline-based, 122
discipline development, 29, 30
domain-based, 31, 35, 122, 147
in RLGs, 99, 104-106, 111
policy analysis, 42, 184

National Health Service (NHS), 10, 20, 40, 41, 49, 55, 80, 84, 86, 88, 109, 189, 191-200, 203, 205
National Institute for Health and Clinical Excellence (NICE), 199
National Perinatal Epidemiology Unit, 98
National Service Frameworks, 197
Natural Environment Research Council, 15
natural sciences, 32, 33
needs, research, 15, 19, 20, 67, 103, 104, 160, 168, 174, 177, 189, 207
social, 20, 28, 81, 86

negotiation, 11-13, 16, 17, 19, 23, 28, 29, 47, 51, 76, 110, 160, 204
collaboration model, 17, 209
exchange model, 11, 12, 204
neutrality, government, 47
New Public Management, 48, 196
NHS R&D Programme, 193, 194, 196, 198, 200, 205, 220
nursing, 10, 17, 30, 62, 93, 94, 99, 100, 109, 121, 126, 132, 140, 147, 162, 184, 186, 189
RLGs, 94, 97, 99, 100, 109, 189

objectivity concept, 101, 105, 122
Office of the Chief Scientist (OCS), 56, 61n, 97n, 103, 104, 112, 129, 131, 132, 141, 153, 156, 160, 167, 170, 181, 184, 213
commissioning role, 103, 153, 160
coordinating role, 133, 153, 160
see also research management system
Office of Population Censuses and Surveys (OPCS), 118, 189
organisational structures, 13, 36, 91

Panel on Health Studies (SSRC), 88
Panel on Medical Research (PMR), 18, 19, 53, 61, 77, 168, 204, 214
MRC and, 19, 73, 77
role (DHSS policy), 53
participative systems, 42
perinatal mortality (RLG project), 174
personal social services, 17, 19, 52, 61, 64, 109, 110, 115, 120-123, 127, 131, 134, 146, 161, 170, 177, 178, 213
SGC applications, 109, 110
social care seminar, 67, 102
see also Health and Personal Social Services
Personal Social Services Research Group, 19, 20, 61, 73
planning, 41-44, 50, 55, 59, 64, 67, 68, 75, 83, 95, 130, 135, 166, 169, 178, 183, 186, 188, 208
government policy, 39
programming-budgeting-system, 43
RLG research, 62, 69, 169
Plowden Report (1961), 42, 43
pluralism, 40, 47, 48, 51, 162, 209
policy
analysis, 42, 83, 178, 184
divisions (customer role), 5, 13, 177, 178
impact of research, 165-175
issues/problems, 12, 14, 31, 59, 63, 65, 70, 71, 95-97, 103, 108, 121, 147, 148, 162, 182, 183, 185-187, 207, 209
macro-scientific, 12, 53, 61, 71, 72
oriented research, 31, 124
planning, *see* planning
processes (government), 49, 50
related research, 44, 87, 115, 182, 206
RLG role, 19, 112, 119, 167, 170, 174, 177
system (DHSS), 42, 58, 76, 124, 184, 216
policy relevance, 54, 87, 98, 104, 109, 131, 145, 146, 148, 153, 158, 160-162, 168
assessment of, 153, 160
authority allocation and, 46, 161
policy-maker judgements, 146, 148
scientific merit and, 53, 54, 131, 136, 145, 146, 148, 158, 160, 161
policy-makers, 3, 13, 15, 16, 19, 20, 31, 33-35, 43, 47, 50, 53, 68-70, 74-76, 82, 86-89, 91, 93, 95-97, 100, 102-105, 107, 108, 112, 124, 130, 137, 143, 144, 146-148, 153-163, 165-169, 173-175, 178, 184-187, 191, 195, 197, 204-207, 209-211, 214, 216

exchange relationships, 52, 106, 206
policy-making, 1, 15, 35, 41, 44, 46, 49, 52, 57, 63, 64, 68, 75, 88, 160, 177, 188, 189, 191, 195, 197, 199, 203, 206, 216
see also collaboration model
Policy Research Programme (PRP), 194, 198
Policy Strategy Unit, 183
politicians, 44, 45, 47
politics, 2, 42, 48, 51, 106, 125, 126, 136, 158
polycentralist planning, 41
positivism, 55, 121, 145
poverty programme (US), 33
power, 7-13, 25-27, 29, 35-37, 40-42, 45, 47, 50-52, 54, 56, 61, 62, 73, 86, 89, 90, 104-106, 108, 111, 112, 115, 120, 125, 126, 131, 133, 143, 149-151, 160, 161, 180, 185, 191, 193, 202, 203, 205-207, 222, 223
-authority relationship, 7, 8, 52
differential, of scientists, 206
imbalance 11, 52, 205
pressure groups, 9, 42, 46, 56, 173, 209
preventive medicine, 63, 66, 67, 74, 84, 102, 200
problem-solving, 29, 31, 181
programme, rolling, 98
programme analysis review (PAR), 44
programme budgeting (in PPBS), 30-31
project commissioning, *see* Department of Health and Social Security commissioning system
psychiatry, 93, 103, 105, 119, 121, 122, 184
Public Accounts Committee (PAC), 24, 80, 83, 84, 116, 125, 215
public engagement, 3, 49, 192, 196, 223
public expenditure, 42, 43, 193
rationality, 10, 13, 27, 39-42, 44, 55, 56
receptor bodies, 3, 199
regional liaison divisions, 58
research
action, 34, 100, 101, 111, 124, 139, 145
applied, *see* applied research
biomedical, *see* biomedical research
commissioning, *see* Department of Health and Social Security commissioning system
Development Controller, 1, 2, 4, 13
dissemination, 100, 170, 172, 192
domain-based, 31, 35, 147
evaluative, 33, 72, 100, 101, 112
function of, 159, 165
impact of, 34, 35, 165, 175, 210 *passim*
multidisciplinary, *see* multidisciplinary research
needs, *see* needs, research
planning, *see* planning
policy and, *see* policy relevance
policy development and, 78
reception of, 170
resources, 67, 82, 98-100, 115, 117, 215
Research Assessment Exercise, 201, 204
research council system, 14
brokerage in, 182, 205
see also individual councils
Research Initiatives Board (SSRC), 67, 87
Research Liaison Groups (RLGs), 19, 20, 43, 53, 62, 95, 97, 99-107, 112, 119, 120, 129, 134, 153-155, 167-170, 174, 189, 185, 187, 189, 192, 213, 214
boundary problems, 102, 103
CSRC role, 48-52, 62, 93, 95, 169, 205

customer role, 19, 93, 96, 104, 155-157, 167, 170, 177, 178, 180, 187
dissemination, 99, 100, 102, 105, 106, 170, 172
exchange relationships, 106-108
methods, 68-71
objectives/effectiveness, 65, 70, 94-98, 101, 105, 179
policy relevance role, 168
range/terms of reference, 93, 94
research strategy, 101, 104, 106, 155, 156, 168
research units, 67, 94, 99, 119, 120, 153, 154
researcher resources, 99
scientific accountability, 20, 103
Small Grants Committee and, 20, 53, 63, 93, 108-112
research management system, 4, 55, 93, 102, 155, 160, 213
dissemination practices, 102
liaison officers, 186, 205
review of, 93, 160
scientific advisers and, 93, 94, 214
Small Grants Committee and, 93, 108
Research Resources Panel (CSRC), 68
Research Strategy Committee (DHSS), 21, 182
research units, DHSS-funded, 20, 115, 118, 126, 129, 131, 170, 214, 216
coherence and power, 125-128
constitution and contracts, 116
customer review, 5, 143, 153-163
HPSS research, 120, 121
intra-unit relationships, 123-124
liaison officer role, 129-133, 186-188
networks, 124-125
review of, 20, 54, 153
social care seminar, 67, 102
theoretical conflicts, 121-123
visits, *see* Chief Scientist visits
research units, MRC, 182
researcher resources, 99
Resource Allocation Working Party (RAWP), 178
resources, 7-9, 11, 16, 17, 26, 36, 40, 41, 43, 51, 52, 65-68, 72-74, 82, 96-100, 115, 117, 137, 138, 150, 157, 166, 168, 172, 180-182, 186, 193, 196, 197, 201, 206, 211
allocation of, 26, 46, 67, 82, 161, 196
Rothschild Report (1971), 2, 4, 10, 12, 14, 36, 44, 59, 83, 88, 181, 191, 204
DHSS version, 10, 12, 36, 116
reactions, 15, 16
Rothschild Report (1982), 116, 118
science
authority of, 24, 25, 35, 108, 131, 160, 161, 206
cognitive structures, 26, 191
convergence/divergence, 8-10, 202
exchange theory, 11, 52
externalist model, 27, 28, 36
-government (models), 4, 7, 8, 21, 108, 206, 207
influence of, 13, 15, 161
internalist model, 7, 8, 23, 25, 27
multimodality of, 5, 198, 202, 203
social function, 25
steerage, *see* steerage
scientific accountability, 20, 103, 131, 135, 169, 181
scientific advisers, 17-20, 44, 65, 68, 93, 97, 100, 104-107, 112, 129-131, 133, 134, 144, 146, 135, 155-158, 162, 171, 180-182, 184, 185, 205, 211
Chief Scientist visit, 112, 131, 133, 181, 182
exchange relationships, 104
RLG, 93, 94, 101, 104, 105, 107, 133, 155, 156
scientific judgement, 24, 71, 76, 85, 144, 158, 161, 185
scientific merit, 23, 53, 109, 110, 128, 131, 134, 136, 137, 140, 144-146, 148, 157-161

adversarial mode and, 138, 148-151
policy relevance and, 53, 54, 131, 136, 145, 146, 148, 158, 160, 161
scientific standards, 17, 20, 67, 71, 98, 103, 104, 146, 152, 153, 155, 162
RLG, 67, 71, 98, 103
scientists
differential power, 206
exchange relationships, 52, 106, 206
reaction to Rothschild, 14-17
social influences, 13, 27
Scottish Home and Health Department (SHHD), 62, 77n, 78, 79
Seebohm Report, 55
self help (RLG policy), 96
service
Development Group, 57, 58, 74, 178
provision, 67, 74
Service Delivery and Organisation (SDO), 198, 199
situationist interaction, 50
Small Grants Committee (SGC), 19, 20, 53, 61, 63, 86, 93, 108-112, 204, 213, 214
applications, 109, 110
RLGs and, 20, 53, 63, 93, 108-112
social
care, *see* personal social services
disadvantage (RLG research), 102
function of science, 25
needs, 28, 81, 86
research impact (model), 25
security, 17, 55, 56, 63, 64, 70, 109, 178, 192, 197, 215
Social Science Research Council (SSRC), 77, 87-89, 91, 206, 215, 216
social sciences, 17, 28, 32, 33, 62, 121, 126, 161
positivism, 55, 121, 145
social services, 13, 17, 19, 20, 52, 55, 57, 61, 62, 64, 69, 73, 91, 93, 100, 109-111, 115, 120-123, 125, 127, 131, 132, 134, 146, 161, 170, 177, 178, 188, 197, 213, 215
social work, 10, 17, 18, 30, 56, 88, 94, 99, 105, 109, 111, 121-126, 139, 140, 147, 162, 174
sociology, 35, 62, 70, 99, 105, 120, 121, 184, 216
"strong" programme, 27
steerage, 23, 28-30, 36, 81, 89, 112, 207, 208
and finalisation, 112
government, 23, 28, 30, 81, 89
systems model (government), 47 *passim*

technology, values and, 45, 46
triple helix, 2, 226

UK Clinical Research Collaboration (UKCRC), 198, 235
universalism, 24
universities, 2, 14, 82, 90, 115, 126, 127, 201
University Grants Committee (UGC), 16, 39, 91

values, 7, 12, 24, 28, 32, 33, 41, 45-47, 53, 55, 56, 99, 105, 106, 113, 118, 122, 141, 144, 147, 148, 166, 202, 204, 206
authoritative allocation of, 46
technology, and, 45, 46

Welsh Office, 94
Wing panel, 71
women, battered (RLG), 96, 102
World Health Organisation (WHO), 3, 222
working groups, 83, 95, 102-104
working parties, 64, 94, 95, 97, 99, 104, 107, 209

Higher Education Dynamics

1. J. Enders and O. Fulton (eds.): *Higher Education in a Globalising World.* 2002
   ISBN Hb 1-4020-0863-5; Pb 1-4020-0864-3
2. A. Amaral, G.A. Jones and B. Karseth (eds.): *Governing Higher Education: National Perspectives on Institutional Governance.* 2002 ISBN 1-4020-1078-8
3. A. Amaral, V.L. Meek and I.M. Larsen (eds.): *The Higher Education Managerial Revolution?* 2003 ISBN Hb 1-4020-1575-5; Pb 1-4020-1586-0
4. C.W. Barrow, S. Didou-Aupetit and J. Mallea: *Globalisation, Trade Liberalisation, and Higher Education in North America.* 2003 ISBN 1-4020-1791-X
5. S. Schwarz and D.F. Westerheijden (eds.): *Accreditation and Evaluation in the European Higher Education Area.* 2004 ISBN 1-4020-2796-6
6. P. Teixeira, B. Jongbloed, D. Dill and A. Amaral (eds.): *Markets in Higher Education: Rhetoric or Reality?* 2004 ISBN 1-4020-2815-6
7. A. Welch (ed.): *The Professoriate.* Profile of a Profession. 2005 ISBN 1-4020-3382-6
8. Å. Gornitzka, M. Kogan and A. Amaral (eds.): *Reform and Change in Higher Education.* Implementation Policy Analysis. 2005 ISBN 1-4020-3402-4
9. I. Bleiklie and M. Henkel (eds.): *Governing Knowledge.* A Study of Continuity and Change in Higher Education – A Festschrift in Honour of Maurice Kogan. 2005
   ISBN 1-4020-3489-X
10. N. Cloete, P. Maassen, R. Fehnel, T. Moja, T. Gibbon and H. Perold (eds.): *Transformation in Higher Education.* Global Pressures and Local Realities. 2005
    ISBN 1-4020-4005-9
11. M. Kogan, M. Henkel and S. Hanney: *Government and Research.* Thirty Years of Evolution. 2006 ISBN 1-4020-4444-5